Construction Quality Management

Quality management is essential for facilitating the competitiveness of modern-day commercial organisations. Excellence in quality management is a requisite for construction organisations who seek to remain competitive and successful. The challenges presented by competitive construction markets and large projects that are dynamic and complex necessitate the adoption and application of quality management approaches.

This textbook is written in line with the ISO 9001:2008 standard and provides a comprehensive evaluation of quality management systems and tools. Their effectiveness in achieving project objectives is explored, as well as applications in corporate performance enhancement. Both the strategic and operational dimensions of quality assurance are addressed by focusing on providing models of best practice.

The reader is supported throughout by concise and clear explanations and with self-assessment questions. Practical case study examples show how various evaluative-based quality management systems and tools have been applied. Subjects covered include:

- Business objectives – the stakeholder satisfaction methodology;
- Organisational culture and Health and Safety;
- Quality philosophy;
- Evaluation of organisational performance;
- Continuous quality improvement and development of a learning organisation.

The text should prove most useful to students on both undergraduate and postgraduate Construction Management or Construction Project Management courses. It will also prove a valuable resource for practising construction managers and project managers.

Paul Watson is Head of the Department of the Built Environment and Professor of Building Engineering at Sheffield Hallam University, UK

Tim Howarth is a Teaching Fellow and the Director of Student Affairs in the School of the Built and Natural Environment at N

Construction Quality Management

Principles and practice

Paul Watson and Tim Howarth

Routledge
Taylor & Francis Group

LONDON AND NEW YORK

This edition published 2011
by Spon Press

Published 2016 by Routledge
2 Park Square, Milton Park, Abingdon, Oxon OX14 4RN

Simultaneously published in the USA and Canada by
Spon Press
711 Third Avenue, New York, NY 10017

Spon Press is an imprint of the Taylor & Francis Group, an informa business

© 2011 Paul Watson and Tim Howarth

Typeset in Goudy by Taylor & Francis Books

British Library Cataloguing in Publication Data
A catalogue record for this book is available from the British Library

Library of Congress Cataloging in Publication Data
Watson, Paul, Dr.
  Construction quality management: principles and practice / Paul
Watson, Tim Howarth.
    p. cm.
  Includes bibliographical references and index.
  1. Building – Quality control. I. Howarth, Tim. II. Title.
TH438.2.W38 2011
690.068'5 – dc22
                      2010034119

ISBN13: 978-0-415-56910-1 (hbk)
ISBN13: 978-0-415-56911-8 (pbk)
ISBN13: 978-0-203-85966-7 (ebk)

Contents

List of figures

List of tables and boxes

Tables

Boxes

Author biographies

Paul Watson Professor Dr Paul Watson is Head of the Department of the Built Environment and Senior Academic at Sheffield Hallam University. Before entering Academia he was engaged on a range of construction projects as a practising construction project manager. However, since becoming an academic he has taught a range of construction management topics at BSc, MSc, and MBA levels. Further, he has supervised and examined at both PhD and DBA levels. He has contributed to the understanding of Quality-related issues by publishing numerous papers and textbooks on the subject. Professor Dr Watson's approach has always been a pragmatic one when it comes to publishing materials, with an emphasis on practical application. Qualifications: MBA, MSc, PhD, FBeng, MCIOB, FHEA, Cert.Ed.

Tim Howarth Tim is a Teaching Fellow in the School of the Built and Natural Environment, Northumbria University, Newcastle upon Tyne. He is a Member of the Chartered Institute of Building and is a Fellow of the Higher Education Academy. He enjoys teaching, research, consultancy and training activities and has authored construction safety management text books and numerous conference and journal papers.

Preface

This book brings together the topics of Quality, Health and Safety, Performance Measurement and Improvement and provides the reader with an accessible overview of key business practices. The structure of the book provides for the facilitation of learning and self assessment.

Learning outcomes are specified at the start of each chapter in order to better enable the reader to readily identify the key principles covered in each chapter. Furthermore, self assessment questions provide the reader with a means to test their understanding of the material presented.

Each chapter includes a useful further reading list, so that the reader can explore and investigate specific topics of interest.

The experience of the authors is reflected in the various chapters, supporting a practical and easy to follow format. The text will prove useful to academics, practising construction industry professionals and students studying for undergraduate or postgraduate degree courses.

Abbreviations

CBPP	Construction Best Practice Programme
CDM	Construction Design and Management Regulations
CIB	Construction Industry Board
CMPS	Centre for Management and Policy Studies
EFQM	European Foundation for Quality Management
HSE	Health and Safe Executive
ILO	International Labour Office
IOSH	Institution of Occupational Safety and Health
ISO	International Organisation for Standards
KPI	Key Performance Indicator
MFAM	Management Functional Assessment Model
OHSAS	Occupational Health and Safety Assessment Series
OHSMS	Occupational Health and Safety Management Systems
QA	Quality Assurance
QMS	Quality Management System
RADAR	Results Approach Deployment Assessment and Review
SEC	Specialist Engineering Contractors
SME's	Small and Medium Sized Enterprises
SMT	Senior Management Team
TQM	Total Quality Management

Introduction

Chapter 1 – An overview of key theorists and quality philosophy

This chapter presents a concise introduction to theories and people that have contributed significantly to the development of the concept and practice of quality management in modern day organisations.

Definitions and notions of quality are presented and the development of quality management practice in modern day organisations is outlined. The contributions of key proponents, theorists and pioneers of quality management are also briefly outlined. Finally, the principles and philosophy of Total Quality Management are explored and the advantages and problematic issues associated with implementing TQM within a commercial context are identified.

Chapter 2 – Measuring project and corporate performance

This chapter explores the advantages to be gained at both project and corporate levels, when construction companies fully engage in the measurement of their performance. It is made clear that improving project and corporate performance requires not only the measurement of performance, but also, the taking of appropriate necessary actions based upon the feed forward of the said measurements.

Various methods of conducting the advocated measurement activities are presented and explained for the reader. The key measurement activities are linked to the critical issue of obtaining stakeholder satisfaction at project and corporate level.

Chapter 3 – ISO 9001:2008

This chapter concentrates on the philosophy and concept of quality assurance encapsulated within the process model of ISO 9001:2008. The various critical aspects of the model and standard are presented and explained for the reader.

The advocated advantages for construction organisations seeking certification to ISO 9001:2008 are established.

Various problematic issues can manifest themselves, and these need to be avoided/addressed, when construction firms commence the application of ISO 9001: 2008, these too are identified. Further suggestions based upon previous practice are made, as to, how to avoid/address the noted problematic issues.

Chapter 4 – The European Foundation for Quality Management Excellence Model (EFQM EM)

The purpose of this chapter is firstly to provide the reader with an introduction to the philosophy and application of Total Quality Management (TQM). Part of the introduction to TQM encapsulates the principles of TQM, along with the advocated advantages of deployment for construction organisations.

Secondly the chapter explores the linkages between TQM and the European Foundation for Quality Management Excellence Model (EFQM EM). The constituent parts of the EFQM EM are presented for the reader, along with the rationale for its deployment in the construction context.

As previously noted in the introduction section for chapter 3, on ISO 9001:2008, there can be problematic issues associated with its application. Similar issues may be encountered when engaging in the implementation of TQM and the EFQM EM. These issues are established for the reader, along with advice on how to best avoid them.

Chapter 5 – Quality and environmental management systems

This chapter introduces the ISO 14000 series of management standards and outlines the requirements placed upon an organisation by ISO 14001: 2004. This standard facilitates the development, support and auditing of environmental management systems within organisations.

Chapter 6 – Developing a learning organisational culture

This chapter presents for the reader a dialogue on the fundamentals of management. These fundamentals are then linked to the concept of Project and Corporate Learning.

However, as noted in the chapter, learning can take place at various levels, and these levels are established and explained for the reader. Further this project and corporate learning activity is placed in the context of project and organisational improvement.

This chapter also incorporates and applies some critical principles, established previously in other chapters of this text book.

One of the key components of this chapter, for the reader, and construction companies, is the presentation and explanation of the Management Functional Assessment Model. The MFA Model is advocated as a means for construction related firms to improve their respective management functions, and hence improve project and corporate performance, all based upon organisational learning.

Chapter 7 – Quality management systems for health and safety in construction

This chapter serves to inform of occupational health and safety management systems and outlines the essential components of such systems for organisations. Advocated benefits and problems associated with occupational health and safety management systems are indicated and differing standards and guidance documents are introduced. The application of a systematic health and safety management approach to construction projects through compliance with the Construction (Design and Management) Regulations 2007 is highlighted. Examples of useful documentation for contributing to the systematic management of health and safety on construction projects are provided at the end of the chapter.

1 An overview of key theorists and quality philosophy

Introduction

This chapter presents a concise introduction to theories and people that have contributed significantly to the development of the concept and practice of quality management in modern day organisations.

Definitions and notions of quality are presented and the development of quality management practice in modern day organisations is outlined. The contributions of key proponents, theorists and pioneers of quality management are also briefly outlined. Finally, the principles and philosophy of Total Quality Management are explored and the advantages and problematic issues associated with implementing TQM within a commercial context are identified.

Learning outcomes

Upon completion of this chapter the reader will be able to demonstrate an understanding of:

- Differing definitions, notions and classifications of 'quality'.
- The contribution of seven key theorists to quality development within organisations.
- Key quality theories that inform and underpin the development and implementation of quality management approaches in modern day organisations.
- Total Quality Management (TQM) and the advocated advantages and problematic issues associated with implementing TQM within a modern day commercial context.

Defining quality

'Quality' is a term that is in common usage within everyday life. It is not unusual to see advertisements that claim 'premium quality', 'purveyors of quality', 'where quality comes first', 'the place where quality counts' and so on.

It is difficult to contest that an association with the term 'quality' offers anything other than positive connotations. The quest to be associated with the notion of 'quality' is key to many modern day organisations. Just what 'quality' means and what the quest to 'achieve quality' entails though is a matter of some debate.

In a search for a definition of 'quality' Reeves and Bednar (1994) point out that 'the definition of quality has yielded inconsistent results ... regardless of the time period or context in which quality is examined, the concept has had multiple and often muddled definitions and has been used to describe a wide variety of phenomena. Continued inquiry and research about quality and quality related issues must be built upon a thorough understanding of differing definitions of the construct.'

When considering 'quality' as a term or concept it soon becomes apparent that it means many different things to many different people. There is quite clearly no one singular, universally accepted definition of 'quality'. The idea or concept of 'quality' is one that is multi-faceted. A survey of the 'definitions of quality' highlights this and is presented in *Table 1.1*. This survey identifies a range of suggested definitions and alternatives that serve to assist

Table 1.1 Definitions of quality

Definition	Source
Definition of quality – a thing is said to have the positive attribute of conformance to specified standards	Shewhart 1931
Quality is a customer determination which is based on the customer's actual experience with the product or service, measured against his or her requirements – stated or unstated, conscious or merely sensed, technically operational or entirely subjective and always representing a moving target in a competitive market	Feigenbaum 1961
Conformance to requirements	Crosby 1979
Quality is (1) product performance which results in customer satisfaction (2) freedom from product deficiencies, which avoids customer dissatisfaction	Juran 1985
Quality: the totality of features and characteristics of a product or service that bears on its ability to meet a stated or implied need	ISO 8402-1986, 'Quality Vocabulary'
Quality is anything which can be improved	Imai 1986
Quality is the loss a product causes to society after being shipped	Taguchi 1986
Quality is the total composite product and service characteristics of marketing, engineering, manufacture and maintenance through which the product in use will meet the expectations of the customer	Feigenbaum 1986
Good quality means a predictable degree of uniformity and dependability at a low cost with a quality suited to the market	Deming 1986
Fitness for use	Juran 1988
Quality is the extent to which the customer or users believe the product or service surpasses their needs and expectations	Gitlow et al. 1989

understanding, use and articulation of the term 'quality' within public and private sector organisations.

It is easy to identify from *Table 1.1* that there is no one singular, universally accepted definition of 'quality'. Indeed attempts to research and define quality within the commercial and organisational contexts of economics, manufacturing, the service industries and strategic and operations management have resulted in, as Garvin (1988) points out, a 'host of competing perspectives each based on a different analytical framework, and employing its own terminology'.

Whilst it can be recognised that definitions of quality are differing, they are not necessarily conflicting or contradictory. Rather, the diversity of definitions underlines the fact that quality is viewed in various ways. This diversity of views and definitions can be problematic though – it can result in confused understanding, articulation and application of the quality concept within public and private sector organisations.

Classification of both the perspectives from which quality is viewed, and the differing definitions of quality, serves to clarify understanding regarding the quality concept. Such classification also serves to underpin and inform both communication and quality management practice. The following section offers an attempt at classifying quality definitions and serves to provide some meaning and structure to the diverse variety of quality definitions.

Classifying the ways of looking at quality

The quality of a product or service can be viewed in purely objective or subjective terms, or in a manner that utilises both objective and subjective evaluation together. *Table 1.2* illustrates the classification of objective and subjective ways of viewing quality.

A research study undertaken by David Garvin (1986) drew upon surveys of 'first-line supervisors' in the USA and Japan and compared practices and attitudes concerning quality. Within this study Garvin identifies five distinct classifications for quality definitions. These five classifications are identified and expanded within *Table 1.3*.

Further to these classifications *Figure 1.1* illustrates Zhang's (2001) 'map of quality perspectives'. This brings together Garvin's five classifications of

Table 1.2 Objective and subjective classifications of quality

Objective quality	Subjective quality
Here the concept of quality is grounded within the precept that the characteristics of a product or service are tangibly measurable and assessable in *absolute* terms such as size, design conformance, durability and performance.	Here the concept of quality is grounded in the *perceived* ability of a product or service to satisfy various needs and aspirations. Here each individual's perceptions can vary regarding the same product or service.

Table 1.3 Five classifications of quality definitions

1	*Transcendent definition of quality*	Here quality is viewed from a perspective of 'abstract properties' – 'I can tell quality when I see it'. Quality is evaluated with knowledge gained from experience and the determination of quality is *subjective* and based upon 'the view of an individual', this view being developed with experience.
2	*Product-based definition of quality*	Here quality is viewed from a perspective of 'desired attributes'. In this context the prescribed features and performance of a product serve to define quality.
3	*User-based definition of quality*	Here quality is viewed from a perspective of 'client / customer satisfaction'. In other words, quality relates to the extent to which client / customer needs and wants are satisfied by the 'fitness for purpose' of the service or product.
4	*Manufacturing-based definition of quality*	Here quality is viewed from a perspective of 'manufacturing compliance'. In other words quality relates to a product's 'conformance to specified requirements'.
5	*Value-based definition of quality*	Here 'quality' is viewed from a perspective of 'economic utility'. In other words, is the service or product value for money? The determination of whether *value* is achieved is a subjective judgement of the client / customer.

Adapted from Garvin

Figure 1.1 Map of quality perspectives

quality definitions and the objective and subjective measurement of quality. In *Figure 1.1* Zhang considers each of Garvin's five quality definition classifications in terms of:

- the extent of the *objective-subjective* determination of each classification of quality definition; and
- the *location* of *where* each quality definition classification is determined (internal or external to an organisation).

The perspectives from which quality may be viewed can be further classified in accordance with an organisation's *product* or *service* function. Product quality and service quality are commonly determined via consideration of differing criteria. Examples of these differing quality criteria are presented in *Tables 1.4* and *1.5*. In *Table 1.4* Garvin (1988) identifies and classifies 8 dimensions of 'product quality'. This can be contrasted with *Table 1.5* where Parasuraman et al. (1988) identify and classify 5 dimensions of 'service quality' in their 'Servqual' model.

Clearly quality is not a singular concept that can be viewed from only one perspective. It has a range of possible definitions and can be seen from a variety of perspectives (subjective, objective, user-based, value-based) and within a variety of contexts (service provision, product manufacture). This range of definitions and ways of looking at and classifying quality developed, evolved and informed quality management practice throughout the twentieth century.

Table 1.4 Dimensions of product quality

Aesthetics	A subjective judgement of a product's look / feel / taste / smell / sound etc.
Conformance	The compliance of the product's characteristics with predetermined physical and performance standards.
Durability	The amount of use of the product before deterioration means replacement is necessary or preferable.
Features	These are distinct properties of the product.
Perceived quality	A subjective assessment of product quality that is influenced by factors such as a product's brand name, image and advertising.
Performance	This is a primary characteristic of concern when considering the operation of the product. For example 'miles per gallon' might be one such characteristic when considering the performance of a car. Clarity of picture and sound may be one such characteristic when considering the performance of a television.
Reliability	This can be described as a product's ability to continue to deliver to required standards over a specified period of time under stated conditions of use.
Serviceability	This can be described as is the speed and ease of maintenance and repair.

Adapted from Garvin (1988)

Table 1.5 Dimensions of service quality

Assurance	Workers' knowledge, courtesy and ability to facilitate trust and confidence.
Empathy	Workers' ability to listen and understand customers and deliver individualised attention.
Reliability	Ability to deliver the promised service dependably and accurately.
Responsiveness	Ability to deliver the service promptly.
Tangibles	These can include the appearance of physical facilities, equipment and personnel.

Adapted from Parasuraman et al. (1988)

The development of quality management practice

The twentieth century saw dramatic growth in production and service industries and the realisation of the global market place and international business organisations.

Post-production inspection predominated quality management practice in the pre-World War II era. Significant changes and developments in quality management theory and practice were seen after World War II. *Figure 1.2* highlights key developments in quality management practice in the twentieth century with the presentation of 'timeline'. *Table 1.6* meanwhile summarises the key attributes and identifying characteristics of the various 'quality movements' of the century.

Key quality theorists and pioneers

A number of 'pioneers' have contributed significantly to the shaping and growth of quality management theory and practice. Each of the following seven 'theorists', 'gurus' or 'pioneers' can be recognised as having distinctively added to an aspect of understanding, development or operation of quality within the management of organisations. The seven notable quality management pioneers are:

- W. Edwards Deming
- Joseph Juran
- Kaoru Ishikawa
- Armand V. Feigenbaum
- Genichi Taguchi
- Philip Crosby
- Masaaki Imai

W. Edwards Deming

William Edwards Deming was born in 1900 and between the years of 1917 and 1928 he enrolled on, and graduated from, a Bachelors degree in

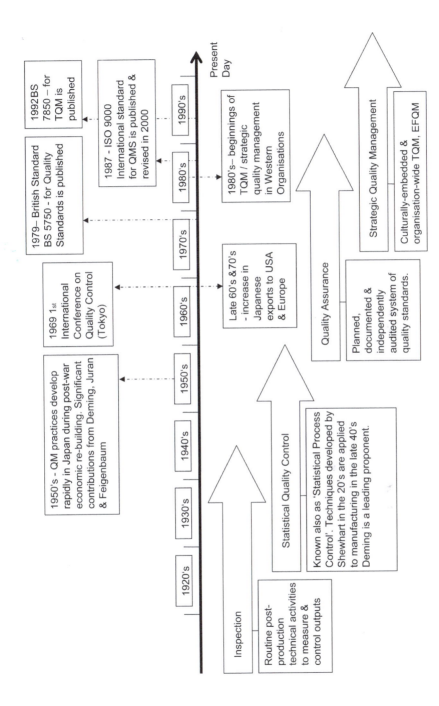

Figure 1.2 Timeline of the key developments in quality management practice

Table 1.6 Key attributes of quality management movements

Identifying Characteristics	Quality Movement			
	Inspection	Statistical Quality Control	Quality Assurance	Strategic Quality Management
Prime Quality Concern	Detection	Control	Co-ordination	Strategic view
Emphasis	Product uniformity	Product uniformity with reduced inspection	The entire production system	The market and consumer needs
Methods used	Measuring	Statistical tools and techniques	Procedural systems	Strategic planning and setting
Role of Quality Professionals	Inspection, Acceptance, Sampling	Trouble shooting and the application of statistical methods	Design of QA system, planning, measurement of performance, audit	Goal setting, education, training and consultation
Responsibility for Quality	Inspection Department	Manufacturing, Engineering Departments	All Departments	Everyone within the organisation
Orientation and Approach	Quality is 'inspected in'	Quality is 'controlled in'	Quality is 'built in'	Quality is 'managed in'

Adapted from Garvin (1988)

Electrical Engineering at the University of Wyoming, attained a Master's degree in Mathematics from the University of Colorado, and gained a doctorate in mathematical physics from Yale University. Further to this Deming took a job at the United States Department of Agriculture where he was responsible for courses in mathematics and statistics.

In 1938 Deming took a position as an advisor in statistical sampling with the United States Government Service's Bureau of Census. Here he applied statistical methods to clerical-operations to establish sampling techniques for the 1940 census. Deming's work realised great improvements in productivity. As a result of his success he was retained in 1942 as a consultant by the War Department. He was later sent to Japan in 1946 by the War Department's Economic and Scientific section to study agricultural production. Here he made connection with the Union of Japanese Scientists and Engineers and was invited to deliver courses in statistical methods to Japanese industry. As a result, Deming delivered lectures in Japan throughout the 1950s on 'statistical methods' as a means of inculcating quality into industry output.

The western world's recognition of Deming's contribution to quality through his work in Japan did not really come about until the 1980's. In this decade he published *Quality, Productivity, and Competitive Position* and

Out of Crisis. Before his passing in 1993 Deming was the recipient of numerous awards and recognitions for his work. These awards include: The Second Order Medal of the Sacred Treasure – Japan's highest accolade to a foreign-national; The American Management Association's Taylor Award; and the National Medal of Technology – presented by President Ronald Reagan in 1987.

Deming's key concepts and contributions to quality theory

Deming's approach to quality is one that strives to 'delight customers'. It is an approach that is concisely portrayed by 'the Joiner Triangle'.

The key concepts and contributions of Deming concern:

- The Plan, Do, Check, Act (PDCA) Cycle – a methodology for problem solving;
- Seven Deadly Diseases of Western Management;
- Fourteen Points for delivering transformation of an organisation for improved efficiency;
- A Seven Point Action Plan for change;
- A System of Profound Knowledge.

Deming places a quality focus upon the causes of variation and variability in an organisation's manufacturing process. In striving to delight the customer, emphasis is put upon the production process, with the deployment of a statistical approach to measure the variability of a given process. For Deming variation is a key factor in poor quality and variation is the result of either a 'Common Cause' or a 'Special Cause'.

Common causes are defined as being systemic and arise from the design or operation of the production system. These causes of variation are viewed

An Obsession with Quality

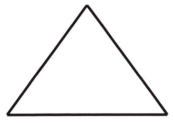

Creation of an environment of 'All One Team'

The deployment of a Scientific Method of approach

Figure 1.3 Fundamentals of total quality leadership

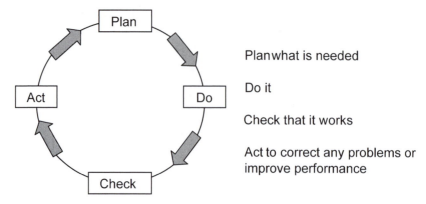

Figure 1.4 Deming's Plan, Do, Check, Act cycle (PDCA cycle)

as being the responsibility of management. Special causes of variation on the other hand are evidenced at a local level by such things as the changing of an operator, shift or machine. These causes of variation are resolved by the giving of attention to each individual cause at the local level.

In Deming's view, management planning is essential if variation, wastage and selling price are to be reduced. The 'Plan, Do, Check, Act' Cycle underlines this necessity and provides a methodology for problem-solving and improvement.

Further to his work in Japan, promoting focus upon variance and the adoption of a systematic approach to problem solving – the PDCA Cycle – Deming identified problems or 'diseases' associated with organisations that required addressing. The 'Seven Deadly Diseases' are identified as:

- Lack of consistency of purpose.
- Emphasis on short-term profit.
- Reliance on performance appraisal and merits.
- Staff mobility.
- Reliance on financial figures.
- Excessive medical costs.
- Excessive legal costs.

Deming's Fourteen Points, for management to enable efficiency within an organisation, were originally presented in his book *Out of Crises*. The fourteen points are applicable to all organisations and are:

- Create constancy of purpose towards improvement of product and service, with the aim to become competitive and to stay in business, and to provide jobs.
- Adopt the new philosophy. We can no longer live with commonly accepted levels of delay, mistakes and defective workmanship.

- Cease dependence on inspection to achieve quality. Eliminate the need for mass inspection by building quality into the product.
- End the practice of awarding business on the basis of price; instead minimise total cost and move towards single suppliers for items.
- Improve constantly and forever the system of production and service to improve quality and productivity and to decrease costs.
- Institute training on the job.
- Institute leadership; supervision should be to help to do a better job; overhaul supervision of management and production workers.
- Institute leadership; supervision should be to help to do a better job; overhaul supervision of management and production workers.
- Drive out fear so that all may work effectively for the organisation.
- Break down barriers between departments.

  — Eliminate slogans, exhortations and targets for the workforce asking for zero defects and new levels of productivity.
  — Eliminate work standards that prescribe quotas. Substitute leadership.

- Remove barriers that rob workers of their right to pride of workmanship (hourly workers; annual or merit rating, management by objective).
- Institute a vigorous education and self-improvement programme.
- Put everybody in the organisation to work to accomplish the transformation. The transformation is everybody's job.

Deming provides organisations pursuing quality via the 'Fourteen Points' and contending with the 'Deadly Diseases' with a seven point action plan for change.

The steps of the seven point action plan are:

- Management struggles over the 14 Points, Deadly Diseases and obstacles and agrees meaning and plans direction.
- Management takes pride and develops courage for the new direction.
- Management explains to the people in the company why change is necessary.
- Divide every company activity into stages, identifying the customer of each stage as the next stage. Continual improvement of methods should take place at each stage, and stages should work together towards quality.
- Start as soon and as quickly as possible to construct an organisation to guide continual quality improvement.
- Everyone can take part in a team to improve the input and output of any stage.
- Embark on construction of organisation for quality.

In his book *The New Economics*, Deming suggests that the prevailing management style of organisations requires a transformation and he proffers a 'System of Profound Knowledge' for this purpose. This system provides 'a map of theory by which to understand the organisations that we work in'. The system of profound knowledge is composed of 4 parts:

- Appreciation for a system
- Knowledge about variation
- Theory of knowledge
- Psychology.

Deming suggested that an individual holding an understanding of the system of profound knowledge would apply it to every kind of relationship with other people. Such a person would therefore:

- Set an example
- Be a good listener, but will not compromise
- Continually teach other people
- Help people to pull away from their current practice and beliefs and move into the new philosophy without a feeling of guilt about the past.

Joseph Juran

Joseph Juran was born in 1904 in Braila in the east of Romania and was raised in Minnesota, USA. In 1924 he graduated with an Electrical Engineering degree from the University of Minnesota and began work as an engineer with the Western Electrical Company in Hawthorne, near Chicago.

In 1926 Western Electrical instituted a statistical quality control process and Juran became a founder employee of this department. He wrote *Statistical Methods Applied to Manufacturing Problems* in 1928, and by 1937 he was the Chief of Industrial Engineering at Western Electric's New York Office.

During the Second World War Juran served a four-year leave of absence from Western. He was employed by the government as an administrator in the Lend-Lease Administration. He improved efficiencies within the department before leaving this Washington post in 1945. He did not return to Western Electrical though, instead he began lecturing, writing and providing consultancy.

Juran joined New York University as a Head of Department and in 1951 he published *Quality Control Handbook*. In 1954 the Union of Japanese Scientists and Engineers invited him to lecture. He continued to lecture and contribute to Japan's economic development throughout the 50s and 60s and later published the lectures he delivered in the 1964 book *Managerial Breakthrough*.

The 1970s brought the publication of *Quality Planning and Analysis*, and the founding of the Juran Institute – a training consultancy for the study of quality management – in 1979. Juran has since continued to actively publish his ideas and on 24 December 2004 he celebrated his 100th birthday.

Juran's key concepts and contributions to quality theory

For Juran quality is concerned with 'fitness for use or purpose', customers are located throughout the production process – both external and internal to an organisation, not simply at the end of the process – and 'breakthrough' to new levels of performance are required if an organisation is to survive and grow. Responsibility for quality is held by management and the awareness and training of management is essential to quality.

'Quality does not happen by accident, it must be planned' is a central view of Juran that is presented in *On Planning for Quality*. Such planning is one component of his developed 'Quality Trilogy'. This trilogy consists of:

- Quality Planning – designing a process that achieves required goals – this requires determining goals, undertaking resource planning, planning implementation and creating a quality plan;
- Quality Control – operating and amending the process so as to achieve optimal effectiveness – monitoring performance, comparing achievements with planned objectives and acting to close any gaps. Here a 'sensor' evaluates the performance of the system and reports to an 'umpire'. The umpire compares actual performance with the required goal and when significant discrepancy exists, the umpire reports to the 'actuator'. The actuator makes adjustments and changes to the system to ensure the achievement of the required goal; and
- Quality Improvement – taking performance to new, superior levels – in terms of satisfying customers, reducing waste, enhancing logistics, improving employee morale and improving profitability.

The quality trilogy places emphasis upon changing and developing the management of quality at the organisation's senior management level.

A 'Quality Planning Road Map' is provided by Juran for an organisation's measured implementation of each planning step. The roadmap details the following necessary steps:

- Identify who the customers are.
- Determine the needs of the identified customers.
- Translate those needs into the organisation's language.
- Develop a product that can respond to those needs.
- Optimise the product features so as to meet the organisation's needs as well as the needs of the customers.
- Develop a process which is able to produce the product.
- Optimise the process.
- Prove that the process can produce the product under operating conditions.
- Transfer the process to operations.

Figure 1.5 Juran's quality trilogy

Further to this, Juran outlines a 'Formula for Results' – this states that an organisation must:

- Establish specific goals to be reached;
- Establish plans for reaching these goals;
- Assign clear responsibility for meeting the goals;
- Base the rewards on results achieved.

Kaoru Ishikawa

Kaoru Ishikawa was born in 1915. He studied at Tokyo University and graduated in 1939 with a Bachelor's degree in Applied Chemistry. In 1947 he became an assistant professor at the University.

Ishikawa was a founding member of the Japanese Union of Scientists and Engineers (JUSE) and attended lectures of Deming and Juran in Japan in the 1950s. By 1960 he had gained a doctorate and became a Professor. He contributed significantly to the development and implementation in the Japanese workplace of numerous tools of quality management, including:

- Quality Circles (the concept was initially published in the journal *Quality Control for the Foreman* in 1962);
- The Ishikawa Graph, also known as the Fishbone Graph, also known as the Cause and Effect Diagram;
- The Seven Tools of Quality of Control – for use by workers within quality circles.

Ishikawa's contribution to quality within Japanese industry further extended with his roles of Chief Executive Director of the Quality Control Headquarters at JUSE and Chairman of the Editorial Committee of *Quality Control for the Foreman*. He wrote and published 2 books before his passing in 1989 – *Guide to Quality Control* and *What is Total Quality Control? The Japanese Way*.

Ishikawa's key concepts and contributions to quality theory

Ishikawa has developed the concept of 'company wide quality control' and is widely regarded for his contribution to the Japanese 'quality circle movement' of the early 1960's and for the development of 'cause and effect', 'Ishikawa' diagrams.

With the deployment of 'quality circles', company-wide quality is advocated. The circles bring an inclusive, accessible and participative approach to quality within an organisation. This bottom-up approach to quality is varied in its application from organisation to organisation but generally consists of circles of between 4 and 12 worker-participants who identify local problems and recommend solutions.

The aims of quality circles are:

- to contribute to the improvement and development of the enterprise;
- to respect human relations and build a happy workshop; and
- to deploy human capabilities fully and draw out infinite potential.

Quality circles deploy seven statistical tools of quality control. These tools are taught to the organisation's employees and consist of:

- Pareto charts (to identify where the big problems are)
- Cause and effect diagrams (to identify what is causing the problems)
- Stratification (to show how the data is made up)
- Check sheets (to illustrate how often it occurs or is done)
- Histograms (to illustrate what variations look like)
- Scatter diagrams (to show relationships)
- Control charts (to identify which variations to control).

One of the tools deployed by the quality circles is the 'cause and effect' diagram, otherwise known as the 'fishbone' diagram or the 'Ishikawa' diagram.

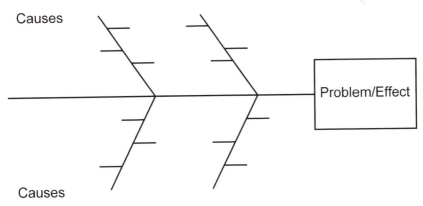

Figure 1.6 Ishikawa / Fishbone / cause and effect diagram

The Ishikawa diagram rather resembles a fishbone and is a systematic tool for investigating the causes of a particular effect and the relationships between cause and effect. The diagram is employed as a tool for identifying opinion regarding the most likely root-cause for a specific, prescribed effect.

Ishikawa claims that when deploying company-wide quality control activities the results are remarkable, in terms of ensuring the quality of industrial products contributing to the company's overall business.

Further to this, the effects of company-wide quality control are offered as:

- Product quality is improved and becomes uniform. Defects are reduced.
- Reliability of goods is improved.
- Cost is reduced.
- Quantity of production is increased, and it becomes possible to make rational production schedules.
- Wasteful work and rework are reduced.
- Technique is established and improved.
- Expenses for inspection and testing are reduced.
- Contracts between vendor and vendee are rationalised.
- The sales market is enlarged.
- Better relationships are established between departments.
- False data and reports are reduced.
- Discussions are carried out more freely and democratically.
- Meetings are operated more smoothly.
- Repairs and installation of equipment and facilities are done more rationally.
- Human relations are improved.

In his 1985 book *What is Total Quality Control?* Ishikawa expands Deming's PDCA Cycle of quality methodology from four steps into six. These steps are:

- Determine goals and targets.
- Determine methods of reaching goals.
- Engage in education and training.
- Implement work.
- Check the effects of implementation.
- Take appropriate action.

Armand V. Feigenbaum

Armand Feigenbaum was born in 1920 and began his working life as an apprentice toolmaker at General Electric. He left General Electric to study for a BA in Industrial Administration at Union College, Schenectady, New York where he graduated in 1942 and attained a PhD at Massachusetts

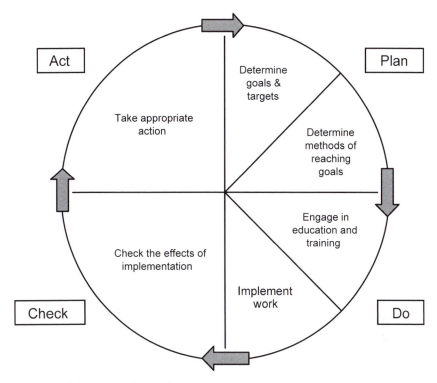

Figure 1.7 Ishikawa's quality cycle

Institute of Technology in 1951. That same year he published his first book entitled *Quality Control: Principles and Practice*.

Quality Control: Principles and Practice was well received in Japan, at a time of industrial regeneration within the country. Feigenbaum's profile in Japan was further facilitated through his role as quality manager and later world-wide director of manufacturing operations with General Electric in the late 1950 and 1960s.

In 1961 Feigenbaum's second book was published, entitled *Total Quality Control*. This book reworked his earlier publication and is recognised as marking the first use of the term 'total quality control'. 1968 saw Feigenbaum leave General Electric to found General Systems Company – a quality management consultancy – with his brother Dr Donald Feigenbaum. Feigenbaum currently remains as President and Chief Executive Officer of General Systems Company, Inc located in Pittsfield, Massachusetts. Further to this, Feigenbaum was the founding chairman of the International Academy for Quality, has twice served as president of the American Society for Quality, was elected to the National Academy of Engineering of the United States in 1992 and has also been the recipient of numerous awards and recognitions for his work with quality.

Feigenbaum's key concepts and contributions to quality theory

Feigenbaum is commonly considered as the originator of 'total quality control' – an approach to quality which advocates quality control as a comprehensive business method and demands quality-mindedness throughout an organisation. For Feigenbaum quality 'has now become an essential element of modern management – it is critical for organisational success and company growth. His own definition of 'total quality control' is provided in his 1961 book of the same name. In his book *Total Quality Control*:

> an effective system for integrating quality development, quality maintenance and quality improvement efforts of the various groups within an organisation, so as to enable production and service at the most economical levels that allow full customer satisfaction.

In attaining the business method of total quality control, three components are necessary:

- Quality leadership.
- Modern quality technology.
- Organisational commitment.

In this total quality context, 'control' is exercised throughout production as a management tool with four steps:

Step 1 Set quality standards.
Step 2 Appraise conformance to the standards.
Step 3 Act when standards are not attained.
Step 4 Plan to make improvements.

The 'Total Quality System' is defined by Feigenbaum as:

> The agreed company-wide and plant-wide operating work structure, documented in effective, integrated technical and managerial procedures, for guiding the co-ordinated actions of the people, the machines and the information of the company and plant in the best and most practical ways to assure customer quality satisfaction and economical costs of quality.

A management tool for measuring the total quality system is provided in the form of 'operating quality costs'. These are categorised as:

- Prevention costs – including quality planning.
- Appraisal costs – including inspection costs.

- Internal failure costs – including scrap and rework.
- External failure costs – including warranty costs and complaints.

More recently 'ten benchmarks for total quality success' have been defined by Feigenbaum. These benchmarks focus the organisation on the customer – both internal and external to the organisation. The benchmarks are:

1 Quality is a company-wide process.
2 Quality is what the customer says it is.
3 Quality and cost are a sum, not a difference.
4 Quality requires both individual and team zealotry.
5 Quality is a way of managing.
6 Quality and innovation are mutually dependent.
7 Quality is an ethic.
8 Quality requires continuous improvement.
9 Quality is the most cost-effective, least capital-intensive route to productivity.
10 Quality is implemented with a total system connected with customers and suppliers.

Genichi Taguchi

Genichi Taguchi was born in 1924. He was a student of textile engineering until he was drafted into the Astronomical Department of the Navigation Institute of the Imperial Japanese Navy between 1942 and 1945. After the war he was employed at the Ministry of Public Health and Welfare and the Institute of Statistical Mathematics.

In 1950 he took a position with the Nippon Telephone and Telegraph Company in the electrical communications laboratory. He became more widely known in this research and development role and served concurrently as a visiting professor at the Indian Statistical Institute between 1954 and 1955.

In 1962 Taguchi gained a Doctorate from Kyushu University, Japan and departed from the electrical communications laboratory of Nippon Telephone and Telegraph Company. He maintained consultancy links and became a Professor of Engineering at Aoyama Gakuin University in Tokyo in 1964. A position he held until 1982, when he became an advisor to the Japanese Standards Institute.

Taguchi continues to serve as Executive Director of the American Supplier Institute – a consulting organisation – and is an advisor to the Japanese Standards Institute.

Taguchi's key concepts and contributions to quality theory

Taguchi has developed an approach to quality that places emphasis upon the product design stage. The cornerstone of his approach is robust design achieved through methodical prototyped reduction of variance in the product. Put another way – Taguchi seeks reduction in variances accepted and tolerated in the production of a unit or item. To Taguchi, the tolerance of acceptable variance in a manufactured product results in what he describes as the 'loss function'.

For Taguchi, any variance from the exact product specification results in what he terms 'quality loss' – even though the product may still be within a traditionally accepted level of variance.

This is illustrated in *Figure 1.8.*

Traditionally product components falling within the level of 'acceptable variance' would be considered useable. Taguchi views this as unacceptable in quality terms, as the failure of a product component to be exactly to target specification will facilitate customer dissatisfaction as the product may perform below its designed optimum. This could result in customer return of the product, customer refusal to purchase another product from the organisation and the customer advising others not to purchase the product. Taguchi clearly identifies the possibility of customer dissatisfaction with a product that is traditionally within tolerable levels of variance.

Taguchi provides a quadratic formula for the 'loss function', this being:

> The formula indicates that as product variation from the specified component target doubles, the quality loss is quadrupled. Dissatisfaction and loss is exponential to distance from the specified quality target.

Taguchi's approach to quality and reduction of the 'loss function' emphasises the need for optimisation of product and process *prior* to manufacture. He provides a methodology for the design of product tests prior to commencing manufacturing. As such, his approach is not one grounded in the pursuit of quality through inspection, instead his developed approach is one of employing 'off-line' quality control.

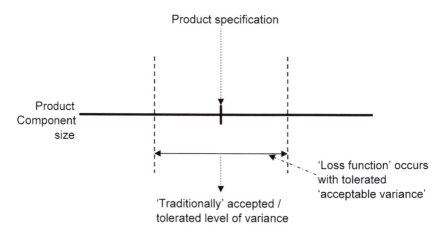

Figure 1.8 Taguchi's intolerance of variance – the loss function

$$L = c(x - T)^2 + k$$

Where:
X is a quality characteristic
T is the specified target
C is the cost of failing to meet the target
k is the minimum loss to society

Figure 1.9 Taguchi's loss function

In pushing quality control back to the design stage Taguchi's methodology advocates the use of prototyping and experimental studies. An 'experimental design procedure' is prescribed for use within the design stage. This procedure provides for the efficient and effective carrying out of simulation experiments with the use of orthogonal arrays (OA) to enable the study of the simultaneous effect of several production process parameters and their interaction.

Reduction of quality loss is achieved via the deployment of a three stage prototyping methodology to the product manufacturing design process. The three stages are:

1 System design – for both the process and product.
2 Parameter design – investigate the optimum combination of process and product parameters. This is done with the objective of reducing sensitivity (and variance and loss factor) in production.
3 Tolerance design – identifies the sensitive components of the design that may give rise to variance in production. Alternatives are considered and tolerance limits are established.

By exercising quality control throughout the product manufacturing design stages, the Taguchi method aims to identify and reduce variance-causing 'noise' factors within the production process and 'optimise' production control factors.

Philip Crosby

Philip Crosby was born in West Virginia in 1926. He served in World War II, graduated from Western Reserve University and further serviced his nation in the Korean War. He began his civilian working life on the production line as a quality professional in 1952.

By 1965 Crosby had become corporate vice president of ITT. He held this position for 14 years. At the end of 1978 he published *Quality is Free*. This was very well received and in 1979 he left ITT and established Philip Crosby Associates Incorporated, a management consultancy. 1984 saw the publication of another commercially successful book *Quality Without Tears*.

Philip Crosby retired from Philip Crosby Associates Incorporated in 1991 and founded a company that delivered lectures and seminars – Career IV Incorporated. In 1996 *Quality is Still Free* was published and a year later he established Philip Crosby Associates II. 1999 saw the publication of his final book *Quality and Me*. Philip Crosby passed away in August 2001.

Crosby's key concepts and contributions to quality theory

Crosby is synonymous with 'zero defects' and a 'do it right first time' approach to quality. He advocates that organisations approach the pursuit of quality in a top down manner, with senior management holding responsibility.

Crosby's Key Concepts on Quality are contained within his:

- Four Absolutes of Quality
- 14 Steps to Quality
- Quality Management Maturity Grid
- 5 Characteristics Essential to becoming an Eternally Successful Organisation

He defines 'Four Absolutes of Quality' and a way for implementing organisational improvement – 'Fourteen Steps to Quality'. The four absolutes of quality are:

- Quality is conformance to requirements.
- The system for quality is prevention.
- The performance standard is zero defects.
- The measurement of quality is the price of non-conformance.

Crosby's Fourteen Steps to Quality are:

1   Make it clear that management is committed to quality.
2   Form quality improvement teams with senior representatives from each department.
3   Measure processes to determine where current and potential quality problems lie.
4   Evaluate the cost of quality and explain its use as a management tool.
5   Raise the quality awareness and personal concern of all employees.
6   Take actions to correct problems identified through previous steps.
7   Establish progress monitoring for the improvement process.
8   Train supervisors to actively carry out their part of the quality improvement programme.

9   Hold a Zero Defects Day to let everyone realise that there has been a change and to reaffirm management commitment.

10 Encourage individuals to establish improvement goals for themselves and their groups.

11 Encourage employees to communicate to management the obstacles they face in attaining their improvement goals.

12 Recognise and appreciate those who participate.

13 Establish quality councils to communicate on a regular basis.

14 Do it all over again to emphasise that the quality improvement programme never ends.

Crosby presents a '**Quality Management Maturity Grid**' in his book *Quality is Free*. This grid serves to provide an organisation with the means to measure its present quality position and is built upon the premise that there are five stages in quality management maturity. These stages are:

1   Uncertainty – management has no knowledge of quality as a positive management tool.

2   Awakening – it is recognised that quality management can help the organisation but no resources are committed.

3   Enlightenment – a decision to introduce a formal quality programme has been made.

4   Wisdom – permanent changes can be made in the organisation.

5   Certainty – quality management is a vital element of organisational management.

According to Crosby there are five characteristics essential to becoming an '**Eternally Successful Organisation**'. These are:

1   People routinely do things right the first time.
2   Change is anticipated and used to advantage.
3   Growth is consistent and profitable.
4   New products and services appear when needed.
5   Everyone is happy to work there.

Masaaki Imai

Masaaki Imai was born in Tokyo in 1930 and graduated with a bachelor's degree from the University of Tokyo in 1955. After undertaking graduate work for the University he founded in 1962 The Cambridge Corporation – a consultancy and executive recruitment organisation.

Imai's prominence in the recruitment field was underlined by his 10-year presidency of the Japanese Federation of Recruiting and Employment Agency Associations. His presidency ended in 1986, the year that Imai established the Kaizen Institute – an organisation for the promotion and support of 'Kaizen' concepts. His first book *Kaizen: The Key to Japan's Competitive Success* was also published in the very same year.

1997 saw the publication of Imai's second book about the Kaizen approach to business. This was titled *Gemba Kaizen: A Commonsense, Low-Cost Approach to Management*. Whilst delivering seminars and lectures, Imai presently continues to run the Kaizen Institute.

Imai's key concepts and contributions to quality theory

Masaaki Imai is a proponent of the Kaizen approach to production and has authored two books concerning the topic. So what is Kaizen? Well it is not a single, distinct quality tool – it is an umbrella concept for a number of practices. It is a philosophy of approach to production developed in Japan. According to Imai, Kaizen means 'improvement, continuing improvement in personal life, home life, social life, and working life. When applied to the production workplace Kaizen means continuing improvement involving everyone, managers and workers alike'. Further to this Imai suggests that Kaizen is the single most important concept in Japanese management – the key to Japanese success.

The deployment of Kaizen is signified by:

- The evolution of processes through gradual continuous improvement rather than by radical change;
- The recognition of the human resource as the prime company resource;
- The quantitative measurement of process performance improvement.

Kaizen is not a prescribed method or quality tool but is a continual striving for an incrementally leaner production process that is driven by workplace teams and has improved production process documentation. The management function of Kaizen is considered to be made up of two elements – maintenance and improvement. The 'maintenance' element of the Kaizen approach to production management concerns sustaining current standards through the deployment of policies, rules and standard operating procedures. The 'improvement' element is concerned with incremental improvement – Imai views improvements as being either gradual 'kaizen improvements' or abrupt 'innovations'.

Key features of Kaizen are:

- The empowerment of employees through the use of Kaizen support groups, quality circles and education. People are at the very heart of Kaizen.
- The use of a range of quality tools by employees – including Deming's Cycle and Ishikawa's seven tools – Pareto charts, cause and effect diagrams, stratification, check sheets, histograms, scatter diagrams and control charts.

- The standardisation of workplace processes.
- The undertaking of 'good housekeeping' within the workplace by everyone – using a system known as the '5 S' – to ensure effective workplace organisation and continuous incremental improvement. This system involves:

  - 'seiri' – 'sorting out' what is not needed around the individual's workplace,
  - 'seiton' – 'systematically arranging' what is to be kept,
  - 'seiso' – 'scrubbing spick and span' everything that remains,
  - 'seiketsu' – 'spreading and standardising' the routine to others, and
  - 'shitsuke' – 'self-discipline' of establishing a routine schedule for the carrying out of the '5 S'.

- The elimination of 'muda' – this being waste caused by any non-value activity. Eliminating 'muda' creates a leaner, just-in-time production process.

With regard to the Kaizen approach to business practice Imai defines the role of management as being such that they must 'go to gemba', the word 'gemba' being a Japanese word for 'real place' – the place where the real action happens. In the terms of business activity this place is seen as anywhere where value-adding activities to satisfy the customer are carried out. In the general sense gemba might be where development, production or sales activity takes place. 5 principles of gemba-management are presented by Imai:

- When a trouble happens (something abnormal), go to gemba first.
- Check with 'gembutsu' (machines, tools, rejects and customer complaints).
- Take temporary counter-measures on the spot.
- Find out the root cause.
- Standardise for prevention of recurrence.

A current leading advocate of Kaizen is Toyota – with a lean, just-in-time approach to production, the creation of a continuous learning culture and the expansion of the employee role.

The Total Quality Management (TQM) approach

The contribution of the seven identified key proponents of quality is significant and each has furthered understanding, development and application of quality management within modern day organisations. All have contributed in differing ways to the post World War II 'quality revolution' and to various ways of thinking about quality. A key feature of the Western quality revolution of the later part of the twentieth century was the development of a strategic approach to quality management. This approach was labelled 'Total Quality Management'.

Defining Total Quality Management (TQM)

Total Quality Management (TQM) is a management approach, centred on quality, based on the participation of all members and aiming at long-term success through customer satisfaction (BSI 1995, cited by McCabe 2001).

However, since the concept of quality consists of both qualitative and quantitative aspects, quality cannot be directly measured; its assessment contains an element of subjectivity.

Smith (1993) has established four specific factors that impact upon the function of 'quality assessment'. They are the determination of user needs, the identification of entity attributes, assessing the entity's merit on each of the associated attributes, and consolidating the established partial scores into a final judgement of quality.

A further aspect requiring consideration with regard to quality is the concept of distinguishing between 'quality' and 'grade'. 'Grade' may be defined as a category or rank given to entities having the same functional use but different technical characteristics. It is worth noting that low quality is usually a problem, but low grade may not be. *Table 1.7* provides an illustration of 'grade' and 'quality' (Project Management Institute 2001).

Quality Definitions for TQM

1. **Quality Policy**: policy includes the quality objectives, level of quality required by the organisation, and the allocated roles for organisational employees in carrying out policy and ensuring quality. It shall be supported and implemented by senior management.
2. **Quality Objectives**: objectives are a critical component of the quality policy. For example a quality objective could be to ensure the training of all employees on the quality policy and objectives of the host organisation.
3. **Quality Assurance**: Kerzner (2001) defined Quality Assurance as a 'collective term for the formal activities and managerial processes that are planned and undertaken in an attempt to ensure that products and services are delivered at the required quality level'.
4. **Quality Control**: Quality Control can be defined as *'a collective term for activities and techniques, within the process, that are intended to create specific quality characteristics'*. In other words, it will assure that the organisation's quality objectives are being met, by using certain techniques such as continually monitoring processes and statistical process control (Kerzner 2001).
5. **Quality Audit**: Kerzner (2001) opined that it is *'an independent evaluation performed by qualified personnel that ensures that the project is conforming to the project's quality requirements and is following the established quality producers and policies'*.
6. **Quality Plan**: project team members will create a specific quality plan for the delivery of a specific project. The plan should contain the key elements/activities of the project and explain in sufficient detail exactly how they are to be delivered and assured.

Table 1.7 Grade and quality defined

Software Product (1)	High quality (no obvious bugs, readable manual) and low grade
Software Product (2)	Low quality (many bugs, poorly organised user documentation) and high grade (numerous features)

Smith (1993) addressed the specific issue of how to define quality within a framework of TQM. He suggested it should incorporate two main features:

- Quality should be taken as the goodness or excellence of organisational products, processes, structures and other entities that an organisation consists of.
- Quality should be assessed against accepted standards of merit and focus on the requirements of stakeholders.

Accordingly, Quality for TQM purposes can be defined as:

> Quality is the goodness or excellence of any product, process, structure or other thing that an organisation consists of or creates. It is assessed against accepted standards of merit for such things and against the interests/needs of products, consumers and other stakeholders.
>
> (Smith 1993)

Erridge et al. (1998) advocated that the main incentive behind adopting quality initiatives in the UK public sector was attributable to the success of TQM in the private sector. Furthermore, government initiatives have encouraged the application of quality management, noting that it could increase the standards of services being offered by public sector organisations without any corresponding increase in public spending.

Moreover, in 1999, the UK government published a white paper entitled "Modernising Government" which consisted of five commitments:-

- To deliver policies that achieve outcomes that matter;
- to deliver responsive public services that meet the needs of citizens, not the convenience of the service provider;
- to deliver efficient, high-quality services and not tolerate mediocrity in service provision;
- to be proactive in the use of new technology in order to meet the needs of citizens and business, and not trail behind technological developments;
- to value public service and not denigrate it.

The above publication was generated by the Modernising Government Quality Schemes Task Force, which was established in January 1999 by the Cabinet Office. The Cabinet Office led the Task Force with members drawn from across government and organisations managing quality schemes, such as the British Quality Foundation and Investors in People UK.

Another important step was taken towards the improvement of service provision in public sector organisations by establishing the "Centre for Management and Policy Studies" (CMPS). Its purpose is to work with government departments and others in a drive to modernise government. It has been working with the Civil Service College, in order to assess and review the training of civil servants and this has resulted in the creation of new programmes.

Furthermore, Capon et al. (1995) briefly summarised the history of how quality was viewed during the last century. Quality has been measured by the percentage of failures. Then, as prevention and Quality Assurance (QA) became more prevalent, SPC and procedural audits provided key measures of its effectiveness. In the 1980s, with cultural change encouraged in a drive for continuous improvement in manufacturing and service provision, employee attitude surveys became popular. In the 1990s, the holistic nature of TQM was adopted, which has encouraged customers', shareholders' and competitors' reactions to become part of the assessment process when assessing the effectiveness of a TQM venture.

Most projects have the conflicting criteria of time, cost and quality. People have differing expectations of quality and these expectations compete with the criteria of cost and time. On this subject Woodward (1997) suggests that time, cost and quality are the prerequisite objectives of any project, these are not 'compatible' and compromises must be made to try and find the best criteria that fit a particular situation (Woodward 1997). However, projects, be they manufacturing or service centred are delivered by people. Projects that involve human concerns will raise particular sensitive issues and how these issues are dealt with will affect the project's outcome and hence the consideration given to the criteria of time, cost and quality. Further McGeorge and Palmer (2002) recommended three approaches for considering the relationship between quality and cost:

1   'Higher quality means higher cost'
    If a higher standard of quality is required this usually results in higher costs. In such cases the benefits obtained should be at least equal to the additional cost paid to get the high standard of quality.
2   'The cost of improving quality is less than the resulting savings'
    Sometimes during the design stage extra costs have to be incurred in order to improve project quality, this should result in less costs being incurred over the life of the project.
3   'Right-first-time approach'
    The costs associated with 'not getting it right first time' are higher than the associated costs of 'getting it right first time', thus investment in getting it right first time is a worthwhile investment.

The above provides some criteria for engaging in the decision making process associated with the time, cost and quality dilemma.

Haigh and Morris (2001) noted that the most common difficulty organisations encounter when embarking on the deployment of Total Quality Management (TQM), are the various definitions of TQM. The most recognised and widespread definition of TQM is provided by ISO 8402 (BSI 1995) (formally BS 4778 part 3 1991) which is '*a management philosophy embracing all activities through which the needs and expectations of the customer and the community, and the objectives of the organisation are satisfied in the most efficient way by maximising the potential of all employees in a continuing drive for improvement*'. While the British Quality Foundation (1998) categorises the definitions of quality into three different types as follows:

- **'Soft aspects'** which are concerned with culture, customer orientation, teamwork, and employee participation.
- **'Hard aspects'** which are mainly technical aspects such as methods, control of work, and procedures.
- **'Soft and Hard aspects'** address both the technical and humanistic aspects of TQM.

Accordingly, Haigh and Morris (2001) tried to simplify the concept of TQM:

- TQM is a total system of quality improvements with decision-making based on facts rather than feeling.
- TQM is not only about the quality of the specific product or service but it is also about everything an organisation does internally to achieve continuous performance improvement.
- TQM assumes that quality is the outcome of all activities that take place within an organisation, in which all functions and all employees have to participate in the improvement process. In other words an organisation requires both Quality Systems and a Quality Culture.
- TQM is a way of managing an organisation so that every job and every process is carried out right first time every time. The key to achieving sustainable quality improvement is through the adoption of TQM principles.

In brief, TQM focuses on a systematic approach to optimally utilise all activities in order to achieve improvements. Therefore '*The key aspects of TQM are the prevention of defects and an emphasis on quality in design. TQM is the totally integrated effort for gaining a competitive advantage by continuously improving every facet of an organisation's activities*' (Ho 1999).

To simplify the meaning of TQM, Ho (1999) proposed a definition for each word that constitutes TQM:

Total: Everyone associated with the firm is involved in continuous improvement (including its customers and suppliers if feasible);

Quality: Customers' expressed and implied requirements are met fully;

Management: Executives are fully committed. Ideally, everyone in the organisation should be committed.

Griffith et al. (2000) summarised the whole process of TQM as follows *'TQM is a philosophy for achieving a never ending improvement through people'*. Clearly this statement defines the two essential key factors needed during the process of implementing TQM:

- Continuous improvement
- People

Haigh and Morris (2001) advocated that organisations need both "quality systems" and "quality culture". In addition, they added that the transition to sustainable quality improvement cannot be achieved except after embracing and implementing the TQM principles. They further advocate that the works of quality gurus such as Deming, Juran and Crosby could form a basis for understanding the principles of TQM. Furthermore the advocated 'best way' to utilise the works of quality gurus is for an organisation beginning its TQM programme to adopt one of the gurus' works. However, during the implementation process the host organisation should develop its own model, one that better fits its specific criteria.

Deploying Total Quality Management (TQM)

First it should be noted that the deployment of TQM should be predicted on realistic expectations about *"What TQM can deliver even those TQM implementations that have delivered good results but below expectations may be perceived to be failures"* (Hendricks and Singhal 2001).

Thus realistic target setting is an activity that Senior Managers must give consideration to, and communicate to staff.

The principles of TQM embrace the concept of customer/supplier relationships existing both within companies (between one person or department and another) and between companies. At each of these interfaces there must be a dedication to meeting the stated requirement with perfection being the only accepted objective. Issues to be addressed as principles of TQM are: leadership, commitment, total customer satisfaction, continuous improvement, total involvement, training and education, ownership, reward and recognition, error prevention, co-operation and teamwork (Oakland 1993).

Most if not all construction organisations seek to implement TQM as a valid means of obtaining for their respective organisations a truly sustainable competitive advantage. Competitive advantage has been defined as:

> an advantage your competitors do not have.

(Hardy 1983)

Powell (1995) showed that under the resources model, success derives from utilising economically valuable resources that other firms cannot imitate, and for which no equivalent substitute exists. Quality Management can improve a firm's competitiveness through co-operation. Cherkasky (1992) stated that when quality concepts are applied to every decision, transaction and business process, quality becomes a competitive weapon. However, processes which have the greatest impact on customer satisfaction would have to be targeted for improvement and only market research would identify the "key customer drivers" or those products and service attributes of greatest concern to customers.

Chapman et al. (1997) argued that although there was a perception that a quality-driven strategic advantage had a direct link with increased business performance, the latter had been difficult to achieve without the development and implementation of a TQM philosophy. Chien et al. (1999) highlighted the factors related to competitive advantage; they identified the following sub-headings of Manufacturing, Marketing, R&D and Engineering and Management.

Fahy (1996) contended that competitive advantage for service firms lay in the unique resources and capabilities possessed by the firm. Not all resources or capabilities are a source of competitive advantage. Only those that meet the stringent conditions of value, rareness, immobility and barriers to imitation are true sources. The actual sources of competitive advantage are likely to vary depending on the nature of the service, the particular traits of the firm, the nature of the industry and the country of origin. *Figure 1.10* provides a pictorial representation of the interrelationship of the sources of competitive advantage, positions of competitive advantage and performance outcomes (Day 1990).

Fahy (1996) concluded that service firms should seek to identify the skills and resources they possess and that they must satisfy the above criteria in order to realise a sustainable competitive advantage.

The linkages between a quality strategy and competitive advantage, though pursued by construction organisations, are very rarely understood within the organisations involved. Improving competitiveness is one of the primary goals of quality management (Rao et al. 1997). Therefore firms need to identify their sources of competitive advantage in order to fully satisfy their clientele.

The linkage between TQM and competitive advantage

From research conducted it has been seen that organisations implementing TQM demonstrate improvements in their efficiency and effectiveness (Chileshe 1996). In the words of one organisation, pursuing TQM had resulted in them being asked to tender for more contracts.

During the onset of the 1992 recession in Australia major problems arose. Hoffman (1992) identified these as the economy, government reforms, interest

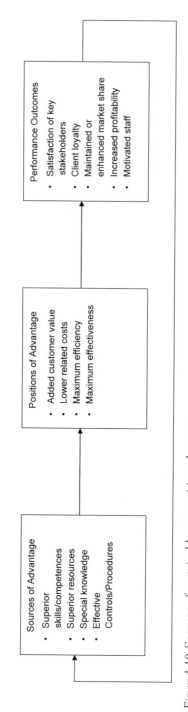

Figure 1.10 Sources of sustainable competitive advantage

rates and the lending market, shortage of labour, and lobby group pressures. Hoffman further pointed out that while some companies had 'gone to the wall' others had profited, improved and gained in strength during the same period. His study dealt with the positive elements common to those companies that had profited. The common element was TQM. This verified the hypothesis that TQM improved the efficiency and effectiveness of an organisation. Oakland (1993), as cited by Ghobadian and Gallear (1996), reported the results of a study that compared the performance of 29 companies practising TQM, along seven key financial indicators for a 5-year period, with a corresponding industry median. The study showed that the performance of all the companies that had adopted TQM exceeded their respective industry's median performance level.

Porter's (1980) framework for the analysis of competition in specific industries showed that an industry had a high level of competitive rivalry when:

- it is easy to enter the market place;
- both buyers and suppliers had a bargaining power;
- there is a threat of substitute products/services entering the market place.

Although Porter's analysis of competitive forces did not specifically address TQM, it does provide a framework for establishing the role that TQM could play in a company's competitive strategy.

The structural implications of TQM for service and manufacturing organisations can be addressed by asking the following key questions:

a Can TQM be utilised to build barriers against new entrants to the industry?

The barriers of entry are largely dependent upon the size of the organisation. Small and medium sized organisations may gain entry into markets, they are however likely to face competition from other smaller firms wishing to become suppliers to larger organisations. This is due to the increasing demand for a higher quality of service from large organisations (Ghobadian and Gallear, 1996). TQM could provide a barrier if clients insisted that it be a pre-requisite before awarding contracts.

b Can TQM change the basis of competition?

Competition is no longer just between firms from the same sector but also now within a global economy from different sectors. Mohrmam et al. (1995) established a positive correlation between various market conditions and the application of TQM practices. The practices included organisational approaches such as quality improvement teams, quality councils, cross-functional planning, self inspection, and direct employee exposure to customers, collaboration with suppliers on quality efforts, just-in-time deliveries and work cells. Various improvement tools such as the use of statistical process control techniques by front-line employees, process simplification and re-engineering; were also evidenced. Measurement systems such as customer satisfaction and cost of quality monitoring also played a vital part. Their studies showed that companies experiencing foreign competition and extreme performance pressures were more likely to use most of the TQM practices, tools and systems. This they suggested provided evidence that competitive pressures had led to the adoption of TQM.

Betts and Ofori (1992) argued that as trade barriers came down, enterprises in each country would face real competition from firms from other countries, even for small projects.

c Can TQM change the balance of power in supplier relationships?

Many companies in the manufacturing industry ensure the quality of their products by requiring suppliers to adopt TQM programs (Powell, 1995 and Mathews and Burati, 1989).

Ghogadian and Gallear (1996) corroborated that small and medium sized enterprises (SMEs) were often suppliers of goods and services to larger organisations and in order to remain competitive, they would have to consider the application of TQM due to the increasing demand for higher quality from the larger organisations.

Moreno-Luzon (1993) identified other factors influencing the application of TQM in small and medium-sized firms as the pressure of costs, increasing competition, and more demanding customers requiring small firms to implement TQM. TQM works by inspiring employees at every level to continuously improve what they do, thus rooting out unnecessary costs. The competitive advantage results from concentrating resources on controlling costs and improving customer service (both internal and external). Dean (1995) purported the challenge to obtaining a sustainable competitive advantage as being able to holistically define the nature of quality and then rigorously implement a form of integrated product and process development (IPPD) which would attain the defined quality.

TQM enables a company to fully identify the extent of its operational activities and focus them on customer satisfaction. Part of this service focus is the provision of a significant reduction in costs through the elimination of poor quality in the overall manufacturing/service process.

The identified characteristics for a TQM company are essential for it to be able to operate both efficiently and effectively in a dynamic and turbulent environment. Firms require variety in their approach and hierarchical authoritarian organisations are poorly equipped to provide such a variety. Only business organisations based on the TQM model with vastly reduced bureaucratic control, a rich array of horizontal communication channels, and in which personnel are given a substantial share of authority to make choices and to develop new ideas, can survive under new global market conditions. Adopting a TQM culture takes a substantial amount of time and effort to achieve.

Problematic issues of TQM deployment

Some TQM proponents maintain that a common error in the application of TQM is the failure to recognise that every company, and environment, is different (Laza and Wheaton 1990, cited by Spencer 1994). Successful deployment is considered to be dependent upon the correct alignment of corporate strategies and operational environments with the culture of the host organisation.

A number of problematic issues are commonly associated with the application of TQM; these include:

- **Insufficient commitment by senior management**. Senior management must instil in all employees of the host organisation a desire to improve the competitiveness of the company. TQM's three vital elements are systems, people, and resources. Successful implementation is dependent upon senior management developing and organising these key elements. Oakland (1993) advocated that '*TQM requires total commitment, which must be extended to all employees at all levels and in all departments*'. Therefore senior management must be fully committed to the implementation processes. This can be evidenced by senior management providing all resources required for the TQM initiative.
- **Incorrect corporate culture**. TQM requires a corporate culture based on trust and a desire to identify problems in order to eliminate them, thus improving production/service process provision. The concept of 'empowerment' is a vital part of the TQM philosophy. However, if a climate of distrust exists between senior management and other parts of the organisation the implementation process is doomed to fail. Organisations must understand that a truly 'morphogenic' change is necessary for TQM success and that a cosmetic 'morphostatic' change will not sustain TQM. Organisational culture clearly influences the manner in which a business organisation operates, and also how employees respond to TQM. As such an organisation's mission statement must recognise the organisation's culture when drawing up tangible targets that are bounded by closed objectives.
- **No formal implementation strategy**. The implementation process should be planned. TQM is a project and therefore requires planning as a project, to treat it as an organisational bolt-on activity will lead to failure. TQM is a means of improving the competitiveness, effectiveness and flexibility of an entire company. Achieving these noted advantages requires them to plan and organise every operational activity at all levels. This process must be part of the strategic implementation development and should not be treated in isolation. Senior management must also understand that the benefits of implementation are not instantaneous; TQM is a long-term corporate investment and having realistic expectations is vital.
- **Too narrowly based training**. The key to a successful TQM implementation is having staff that are competent to execute their allocated

tasks. If employees are empowered to plan and perform work activities it is vital that they also possess the necessary skills and competencies to complete set tasks. A primary function for enterprises seeking to gain a competitive advantage is to implement some form of training initiative which ensures staff have the necessary skills. For example, if staff are to participate in group discussions training in group dynamics and public speaking would be beneficial.

- **Lack of effective communication system**. The life blood of any organisation is communication and the importance of this activity cannot be overemphasised. Within a TQM framework all employees of the company should be able to communicate as necessary, do not forget the concept of 'internal' and 'external' customers with its requirement for effective communication mechanisms. If employees are to become part of the organisational decision-making process they need a means of expressing their views to senior management. Control within any organisation is dependent upon the effectiveness of the communication system's function.
- **Not concentrating on organisational strengths**. TQM is designed to provide a competitive advantage based upon the host firm's strengths. Senior management should not lose sight of the fact that sustained competitive advantages are obtained by implementing strategies that exploit their strengths through responding to environmental opportunities, while neutralising external threats and avoiding internal weaknesses (Barney 1991). The following two standard corporate planning techniques can be utilised; first a Strengths, Weakness, Opportunities, and Threats analysis (SWOT), and, secondly, a Political Legal, Economic, Social Cultural and Technological analysis (PEST).

The approach to implementing TQM

Figure 1.11 presents a model that can be adopted or adapted by organisations in order to assist with the development and deployment of TQM approach.

Management's role in the application of TQM is to create a vision that incorporates TQM as an integral part of the business. Management should further establish organisational policies, structures, and practices consistent with that vision (Fenwick 1991, Ginnodo 1992, Sholter & Hacquebord 1980, cited by Spencer 1994). Managers should be responsible for synthesising all of the different processes and people into a cohesive system (Shores 1992, cited by Spencer 1994).

Thus managers must have a complete understanding of where the company is now, if it wants to deploy TQM and gain the advocated advantages and thus become a 'best practice' organisation.

This will require the company to implement a 'strategic analysis'. Johnson and Scholes (1993) advocate that the strategic analysis should encapsulate

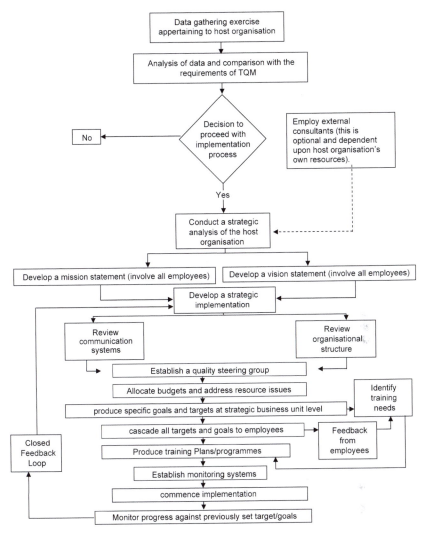

Figure 1.11 A generic model for the implementation of Total Quality Management

'the environment, organisational culture, strategic capability and stakeholder expectations'.

Once the organisation attains a full comprehension of its current state, it then needs to establish where it is in relation to its competitors. This process involves benchmarking various activities of the host organisation.

Advocated solutions to problematic issues of TQM deployment

A number of solutions can be proactively proffered to alleviate problems commonly encountered when deploying a TQM approach. These include:

- Senior management must attain a full understanding of the philosophy and requirements of TQM. Senior management is responsible for establishing a quality-focused organisation;
- A common vision, one that is recognised and shared by all employees of the organisation. This may be accomplished by adopting awareness sessions, customer surveys, benchmarking and common vision workshops;
- Provision of the necessary resources, which include humanistic as well as financial requirements, education and training for quality improvements;
- The development of a holistic deployment strategy that may be based on an incremental process. Senior management must review the quality management systems in order to monitor and maintain progress;
- Design procedural systems relating to work practices. Concentration of organisational effort should be placed on prevention rather than corrective actions;
- Organisations must effectively balance between processes and results. Some organisations focus only on processes and neglect the importance of results.

Factors that influence TQM implementation in public sector organisations

Dewhurst et al. (1999) propose that there are key dimensions that influence TQM implementation within public sector organisations. These dimensions are:

- Top management support: commitment and leadership of top management.
- Customer relationship: culture change, customer involvement, and customer focus.
- Culture change: training, employee empowerment, education and employee relations.
- Supplier relationship: culture change, supplier quality management and supplier involvement.
- Workforce: management and teamwork.
- Employee attitudes and behaviour: employee involvement.
- Product and/or service design process: product/service design and design quality management.
- Process flow management: use of tools and techniques and process management/operating procedures.
- Quality data and reporting: measurement and feedback, quality data and reporting and internal quality information.
- Role of the quality department: it should play an integral part and ensure it leads on the initiative.
- Benchmarking: as a means of self-assessment.

The advantages of applying a TQM approach

TQM can be advocated as a solution for organisations that are underperforming due to their use of traditional organisation structures and management practices whilst operating within a dynamic environment. The implementation of a TQM philosophy can facilitate performance in such organisations.

The advantages of applying a TQM approach are:

- the production of a higher quality product/service through the systematic consideration of client's requirements;
- a reduction in the overall process/time and costs via the minimisation of potential causes of errors and corrective actions;
- increased efficiency and effectiveness of all personnel with activities focused on customer satisfaction;
- improvement in information flow between all participants through team building and pro-active management strategies.

TQM can assist in making effective use of all organisational resources, by developing a culture of continuous improvement. This empowers senior management to maximise their value-added activities and minimise efforts/organisational energy expended on non value-adding activities.

TQM enables companies to fully identify the extent of their operational activities and focus them on customer satisfaction. Part of this service focus is the provision of a significant reduction in costs through the elimination of poor quality in the overall process. This empowers companies to attain a truly sustainable competitive advantage. TQM provides a holistic framework for the operational activities of enterprises. If a firm can overcome the problematic issues of implementation then a sustained competitive advantage is the reward to be gained (Watson and Chileshe 2001).

A tool for assisting management and employees to define and comprehend TQM within their own organisation is the 'European Foundation for Quality Management Excellence Model' (EFQM Model). Details of the nature and application of this model is presented in Chapter 4.

Summary

This chapter has presented various definitions and notions of quality. It has introduced people and theories that have contributed significantly to the development of the concept and practice of quality management in modern day organisations. This chapter has also considered TQM and the advantages and problematic issues associated with implementing a TQM approach within a modern day organisation.

Questions for the reader

Here follows a number of questions related specifically to the information presented within this chapter. Try to attempt each question without reference to the chapter in order to assess how much you have learned. The answers are provided at the end of the book.

Question 1

Define the following terms:

1a) Quality Policy
1b) Quality Objectives
1c) Quality Assurance
1d) Quality Control
1e) Quality Audit
1f) Quality Plan

Question 2

The concept of Total Quality Management has been simplified to four aspects (Haigh and Morris 2001). Identify the four aspects of TQM.

Question 3 – Case study

You have been asked to act as an external consultant for 'Appleyard Innovative Design Solutions'. Appleyard are considering the implementation of a formal TQM system with a view to obtaining externally verified ISO accreditations. Appleyard consider accreditation to be a necessity in order to be placed on tender lists and continuously improve their operations.

As an external consultant, you are requested to prepare and deliver a presentation to the senior partners of Appleyard. The topic of the presentation is 'the benefits of TQM and the associated implementation process'. Prepare notes to facilitate this presentation.

Further reading

Philip Crosby

Crosby, P. (1972) *Situation Management: The Art of Getting Your Own Sweet Way*. McGraw-Hill Education
——(1978) *Quality is Free: The Art of Making Quality Certain*. McGraw-Hill Education
——(1982) *The Art of Getting Your Own Sweet Way*. Higher Education
——(1984) *Quality Without Tears: The Art of Hassle Free Management*. McGraw-Hill Education
——(1996) *Quality is Still Free: Making Quality Certain in Uncertain Times*. McGraw-Hill Education
——(1999) *Quality and Me: Lessons from an Evolving Life*. Jossey Bass Wiley

W. Edwards Deming

Deming, W. E. (1943) *Statistical Adjustments for Data*
——(1950) *Some Theory of Sampling*. Wiley; Chapman & Hall

——(1960) *Sample Designs in Business Research*. John Wiley & Sons Inc
——(1982) *Quality, Productivity, and Competitive Position*. Massachusetts Institute Technology
——(1986) *Out of Crisis*. Massachusetts Institute of Technology
——(1993) *The New Economics: For Industry, Government, Education*. Massachusetts Institute of Technology

Armand V. Feigenbaum

Feigenbaum, A. V. (1951) *Quality Control: Principles and Practice*. New York: McGraw-Hill
——(1961) *Total Quality Control*. New York: McGraw-Hill
Feigenbaum, A. V. & Feigenbaum, D. S. (2003) *The Power of Management Capital*. McGraw-Hill

Masaaki Imai

Imai, Masaaki (1986) *Kaizen: The Key to Japan's Competitive Success*. McGraw-Hill Education
——(1997) *Gemba Kaizen: A Commonsense, Low-Cost Approach to Management*. McGraw-Hill Education

Kaoru Ishikawa

Ishikawa, Kaoru (1984) *Guide to Quality Control*. Asian Productivity Organization, Japan
——(1985) *What is Total Quality Control? The Japanese Way*. Prentice Hall

Joseph Juran

Juran, Joseph (1951) *Quality Control Handbook*. McGraw-Hill
——(1955) *Case Studies in Industrial Management*. McGraw-Hill
——(1964) *Managerial Breakthrough*. McGraw-Hill
——(1982) *Juran on Quality Improvement*. Juran Institute, New York
——(1988) *Juran on Planning for Quality*. The Free Press
——(1998) *Juran on Quality by Design: The New Steps for Planning Quality into Goods and Services: Planning, Setting and Reaching Quality Goals*. Simon & Schuster Inc
Joseph Juran & Frank Gryna (1970) *Quality Planning and Analysis*. McGraw-Hill Education
The Juran Institute (online) – www.juran.com

Genichi Taguchi

Taguchi, Genichi (1986) *Introduction to Quality Engineering: Designing Quality into Products and Processes*. Quality Resources
——(1987) *Methods of Orthogonal Arrays and Linear Graphs*. William C Brown
——(1994) *Taguchi Methods*. William C. Brown
——(2001) *An Introduction to Quality Engineering*. Asian Productivity Organisation, Japan

References

Barney, J., (1991). Firm Resources and Sustained Competitive Advantage, *Journal of Management*, 17(1), pp. 99–120.

Betts, M., and Ofori, G., (1992). Strategic planning for competitive advantage in Construction, *Construction Management and Economics*, 10, pp. 511–32.

British Standards Institution,(1995). *Quality Management and Quality Assurance – Vocabulary*. BS EN ISO 8402 (formally BS 4778: Part 1, 1987/ISO 8402, 1986). London.

British Quality Foundation, (1998). Self Assessment Techniques for Business Excellence. *Identifying Business Opportunities*, London.

Capon, N., Kay, M., and Wood, M., (1995). Measuring the success of a TQM programme. *International Journal of Quality & Reliability Management*, 12(8), pp. 8–22.

Chapman, R. L., Murray, P. C., & Mellor, P., (1997). Strategic Quality Management & Financial Performance Indicators. *International Journal of Quality & Reliability Management*, Vol. 14, No. 4, pp. 432–48.

Cherkasky, S. M., (1992). Total Quality for a Sustainable Competitive Advantage. *Total Quality Management Journal*, August, pp. 4–7.

Chien, T., W., Lin, C., Tan, B., & Lee, W. C., (1999). A Neural Networks-based Approach for Strategic Planning. *Information and Management*, Vol. 35. pp. 357–64.

Chileshe, N., (1996) 'An Investigation into the problematic issues associated with the implementation of Total Quality Management (TQM) within a constructional operational environment and the advocacy of their solutions'. Unpublished MSc dissertation, Sheffield Hallam University.

Day, S, G., (1990). *Market Driven Strategy: Process for Creating Value*, The Free Press, New York, p. 128.

Dean, E. B., (1995). Total Quality Management: from the Perspective of Competitive Advantage, *Total Quality Management Journal*, Pages 1–3.

Dewhurst, F., Martinez-Lorente, A., and Dale, B., (1999). TQM in public organisations: an examination of the issues. *Managing Service Quality*, 9(4), pp. 265–73.

Erridge, A., Fee, R., and Mcllroy, J., (1998). Public sector quality: political project or legitimate goal? *International Journal of Public Sector Management*, 11(5), pp. 341–53.

Fahy, J., (1996). Competitive Advantage in International Services: a Resource-based view, *International Studies of Management & organisation*, Vol. 26, No 2, pp. 24–37.

Garvin, D. (1986), "Quality Problems, Policies and Attitudes in the US and Japan – AN Exploratory Study". *Academy of Management Journal*, Vol 29 No 4, pp. 653–73.

——(1988), *Managing Quality*, The Free Press, New York, NY.

Ghobadian, A., & Gallear, D. N., (1996). Total Quality Management in SMEs. *Omega International Journal of Management Science*, Vol. 24, No. 1, pp. 83–106.

Gitlow, H., Gitlow, S., Oppenheim, H. & Oppenheim, R., (1989) *Tool and Methods for the Improvement of Quality*, Irwin, Homewood, IL.

Griffith, A., Stephenson, P., & Watson, P., (2000). *Management Systems for Construction*. 1st ed., Pearson Education Limited, ISBN 0 582 31927–7.

Haigh, B., and Morris, D., (2001). *Total Quality Management, a case study approach*. 1st ed., Sheffield Hallam University Press.

Hardy, L., (1983). *Successful Business Strategy – How to win the market place*, Kogan Page.

Hendricks, K. B., & Singhal, V. R., (2001). Firm Characteristics, Total Quality Management, and Financial Performance, *Journal of Operations Management* 19, pp. 269–85.

Ho, S., (1999). From TQM to business excellence. *Production Planning and Control,* 10(1), pp. 87–96.

Hoffman, K., (1992). *Improving business performance in the building industry through total quality management,* Conference proceedings for ABIC '92 on efficient & effective construction in the 90's, Australia, Gold Coast.

Johnson, G., and Scholes, K., (1993). *Exploring Corporate Strategy,* London. Prentice Hall ISBN 0-13-297441.

Joiner B. L. (1993) *Fourth Generation Management.* McGraw-Hill Education.

Kerzner, H., (2001). *Project Management: A systems approach to planning, scheduling, and controlling* 7th ed., John Wiley & Sons, Inc.

Matthews, M. F., and Burati, J. L. Jr., (1989). Quality management organisations and techniques, *Source Document 51,* The Construction Industry Institute, Austin, Texas.

McCabe, S., (2001). *Benchmarking in Construction,* Blackwell Science, Oxon, UK, ISBN 0-632-05564-2.

McGeorge, D., & Palmer, A., (2000). *Construction Management: new directions.* 2nd ed., Blackwell Science Ltd.

Mohrmam S, A., Tenkasi, R, V., Lawler III, E, E., and Ledord Jr, G, E., 1995. Total Quality Management: practice and outcomes in the largest US firms, *Employee Relations,* 17(3), pp. 26–41.

Moreno-Luzon, M. D., (1993). Can total quality management make small firms competitive? *Total quality management,* 4(2), pp. 165–81.

Oakland, S.,J., (1993). *Total Quality Management,* Butterworth Heineman, London.

Parasuraman, A., Zeithaml, V. A., and Berry, L. L.: SERVQUAL: A Multiple-Item Scale for Measuring Consumer Perceptions of Service Quality. *Journal of Retailing* 64, 12–40 (1988).

Porter M. E., (1980). The Competitive Advantage of Nations. *Harvard Business Review* March-April.

Powell, T. C., (1995). Total Quality Management as competitive advantage: a review and empirical study. *Strategic Management Journal,* 16, pp. 15–37.

Project Management Institute, (2001). A Guide to the Project Management Body of Knowledge (*PMBOK Guide*). 2000 ed., Project Management Institute, Inc.

Rao, S. S., Ragu-Nathan, T. S., & Slis, L, E., (1997). Does ISO 9000 have an effect on quality management practices? An international empirical study. *Total Quality Management,* Vol 8, No. 6, pp. 335–46.

Reeves C. A. & Bednar D. A. (1994) *Academy of Management Review,* Vol. 19, No. 3, Special Issue: "Total Quality" (Jul., 1994), pp. 419–45.

Shewhart, W. A., (1931) *Economic Control of Quality of manufactured Product,* van Nostrand, New York.

Smith, G., (1993). The meaning of quality. *Total Quality Management,* 4(3), pp. 235–45.

Spencer, B. A., (1994). Models of Organisational and Total Quality Management: A Comparison and Critical Evaluation. *Academy of Management Review* Vol. 19. No. 3. pp. 446–71.

Watson, P., (2000). 'Applying the European Foundation for Quality Management (EFQM) Model', *Journal of the Association of Building Engineers,* Vol 75(4), pp. 18–20.

Watson, P., & Chileshe, N., (2001). The Relationship between Organisational Performance and Total Quality Management within Construction SME's. *Proceedings of CIB World Building Congress*, April 2–6 Wellington, New Zealand, (3), pp. 233–44.

Woodward, J., (1997). *Construction Project Management, Getting it Right First Time.* 1st ed., Thomas Telford Services Ltd.

Zhang, Q. (2001), "Quality Dimensions, Perspectives and Practices: A Mapping Analysis". *International Journal of Quality & Reliability Management*, Vol. 18 No 7, pp. 708–21.

2 Measuring project and corporate performance

This chapter provides a brief introduction to the concepts of stakeholders and self-assessment. These two key concepts are not, and should not be viewed as, mutually exclusive. They are but two sides of the same coin, and the aim of any construction-related enterprise should be to understand who its stakeholders are, and what they require from them. Self-assessment enables a construction company to determine if it is in fact meeting the stated requirements of its stakeholders. Within this chapter are highlighted stakeholder issues, key performance indicators (KPIs) and benchmarking. However, these topics are presented here as 'an introduction to.', because self-assessment in practice is covered in considerable detail in Chapters 3, 4 and 6.

After all, a (construction) business must change to stay ahead or get ahead. If a business does not keep up then its only option is to fall behind (McDonald and Tanner 1996).

Learning outcomes

By the end of this chapter the reader will be able to demonstrate an understanding of:

- The fundamental principles of the stakeholder concept.
- The importance of conducting an organisational and project self-assessment activity.
- The benefits of looking externally as well as internally, when engaging in the self-assessment activity.

Stakeholders

The original concept of Stakeholders related to enterprises and encapsulated groups or individuals who had a financial 'stake' in the corporation. This original concept has now been expanded to incorporate the true concept of 'stakeholder theory'.

Stakeholder theory relates to project and corporate organisational management, business ethics and it addresses both morals and values in managing a construction-related enterprise. A critical and evaluative approach to stakeholders was provided by Freeman (2003). Freeman's work was related to the identification and modelling of groups that could be classified as the stakeholders of companies. Freeman advocated a methodology that can be adopted by management. This methodology addressed the sometimes disparate approach taken by organisations, including construction companies, in identifying stakeholders and their requirements. Freeman put the stakeholder concept in very simplistic terms; it addresses the 'Principle of Who or What really counts' (Freeman 1984).

One needs to give careful consideration not only to who are an organisation's stakeholders, but also to what influence they can exert upon the host company. Mitchell et al. (1997) produced a typology of stakeholders; this work was based on the attributes of power, or the extent that they could influence a company. It is therefore a wise strategy for construction companies to be aware of the stakeholder concept, and the impact its stakeholders could exert upon them.

A construction firm must effectively manage its relationship with its stakeholders, but first it needs to identify its stakeholders. The following provides some examples of typical stakeholders related to their respective interests:

Government	Taxation, Legislation, Lower unemployment
Senior Management	Performance issues, Target setting, Corporate growth & morale, Ethical/green issues & corporate longevity
Non-Managerial Staff	Pay rates, Job security, Morale, Facilities
Unions	Working conditions, Pay rates, Legal requests, Health & Safety issues
Customers	Value for money, Quality, Customer care, Ethical products
Creditors/Suppliers	Liquidity, Timely payments, New contracts, Corporate longevity
Local Communities	Jobs, Local investment, Environmental issues, Ethical practices.

The type of stakeholder, along with their respective interest and influence, will vary from company to company, and industry to industry.

Typical stakeholders

- people who will be affected by an endeavour and can influence it, but who are not directly involved with doing the work. Any group or individual who can affect or who is affected by achievement of a group's objectives;

- any individual or group with an interest in a group's or organisation's success in delivering intended results and in maintaining the viability of the group or the organisation's product and/or service. Stakeholders influence programmes, products and services;
- any organisation, governmental entity, or individual that has a stake in or may be impacted by a given approach to environmental regulation, pollution prevention, energy conservation, etc.;
- a participant in a community mobilisation effort, representing a particular segment of society.

Stakeholders that are classed as 'primary' are those that engage in economic business activity with the host construction company, for example suppliers and subcontractors.

Stakeholders that are classified as 'secondary' are those who – although they do not engage in direct business activity – are affected by or can affect its actions. For example the media, via adverse press related to a construction company or one of its projects.

So the term 'stakeholder' has become more commonly used to mean a person, group or organisation that has a legitimate interest in a project or company. When considering the decision-making processes for enterprises, or government agencies, and non-profit organisations, the concept has been extended to encompass all those who have an interest (or 'stake') in what the enterprise does. This includes not only its vendors, employers and customers, but even members of a community where its offices or projects may affect the local economy or environment.

Stakeholder mapping

The production of a stakeholder map is a very useful activity for construction firms (or any firm) to undertake. The underpinning rationale for the production of a stakeholder map is that:

- It is used to identify all interested parties, both inside and outside the business. Remember any organisation has both internal and external stakeholders.
- It is used to ensure that stakeholder interests are established and catered for.
- It can assist in balancing the needs/interests of the various stakeholders.

Stakeholder content map

A stakeholder map in pictorial form is presented in *Figure 2.1*. The production of a map is a very good starting point when considering the interconnection of internal and external stakeholder relationships.

Stakeholder mapping is an essential activity for construction companies to engage in. It is essential because if a construction company does not know

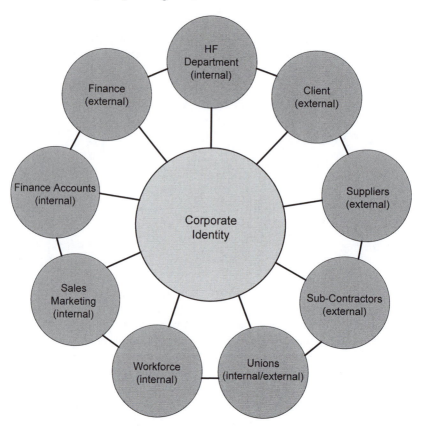

Figure 2.1 Typical example of a stakeholder map

who its stakeholders are, or what they require, how can that construction company possibly meet their requirements?

The development of key performance indicators

Construction companies, or any company for that matter, need to be able to gauge their performance; this is essential if a company is striving for organisational improvement, via organisational learning (see Chapter 6).

Thus the concept of key performance indicators (KPIs) has been embraced by the UK construction industry. The KPIs scheme for construction firms evolved following the report 'Rethinking Construction'. The Construction Industry Task Force prepared this report in 1998 for the Department of the Environment, Transport and the Regions (DETR). The report highlighted a number of areas where the construction industry could improve its performance.

One of the fundamental reasons why the construction industry needed to institute change arose from the statement made in the report:

To drive dramatic performance improvement the task force believes that the construction industry should set clear measurable objectives and then give them focus by adopting qualified targets, milestones and performance indicators.

(Construction Industry Task Force 1998)

The Construction Task Force identified specific targets for improvements:

- Capital Costs reduced by 10 per cent
- Construction time reduced by 10 per cent
- Predictability reduced by 20 per cent
- Defects reduced by 20 per cent
- Accidents reduced by 20 per cent
- Productivity increased by 10 per cent
- Turnover and profits increased by 10 per cent

(McCabe 2001)

Following the production of the 'Rethinking Construction' report, the Construction Best Practice Programme (CBPP) was developed by the DETR, the Construction Industry Board (CIB) and the Movement for Innovation (M4I).

The CBPP was established to institute the challenges that were set to the industry by 'Rethinking Construction'; amongst several other initiatives the KPI scheme was launched in May 1999.

The KPI Working Group (2000) states:

> The purpose of the KPI's is to enable measurement of project and organisational performance throughout the construction industry. This information can then be used for benchmarking purposes, and will be a key component of any organisation's move towards achieving best practice.

Since the formulation of the initial set of KPIs, other indicators have been established; these include People KPIs, Construction Products KPIs and Environmental KPIs.

The Movement for Innovation (M4I) set 10 KPIs, as noted by Cook

> Seven are applied on a project-by-project basis: Construction cost; Construction time; Cost predictability; Time predictability; Defects; Product satisfaction; and Customer satisfaction. Three indicators look at Company performance; Profitability; Productivity; and Safety.

(Cook, 1999)

It is intended that KPIs be used throughout the construction industry by all parties, including clients, designers, consultants, contractors and subcontractors. The 10 KPIs are also covered by:

a set of 'super graphs' that plots the entire industry's average performance for all 10 indicators. There are also five sector-specific sets of graphs covering new-build housing; new-build non-housing; Infrastructure; repair; maintenance housing; and repair, maintenance non-housing.

(Cook 1999)

The use of the graphs assists construction companies in accurately benchmarking themselves against other construction companies.

As construction covers the design, construction and eventually the demolition of a project, this would enable the same project to be measured using the same set of indicators but with the differing parties involved.

The KPI Working Group set out to define the five key stages of a project and these are depicted in *Figure 2.2*.

Construction key performance indicators

The previously stated key performance indicators are known as the 'headline KPIs'. Other performance indicators are in existence and cover the following areas:

- Operational indicators
- Diagnostic indicators

Operational indicators relate to specific aspects of a construction company's activities and should enable senior management to identify and focus on specific areas for improvements. Diagnostic indicators provide information on why certain changes may have occurred in the headline or operational indicators; they are useful in analysing areas for improvement in more detail. This enables a construction company to efficiently and effectively focus its limited organisational resources.

When engaging in this type of activity a construction company must consider the 'law of diminishing returns'. There is always an organisational cost to be paid when engaging in 'Change management activities'.

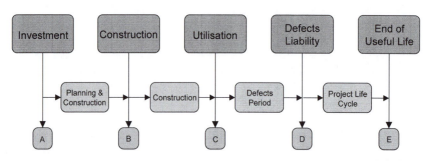

Figure 2.2 Process of commitment

KPIs provide useful tools for engaging in self-assessment activities by enabling the measuring and analysing of the company and/or project, in order to obtain an improved level of performance. The headline indicators provide the main focus of performance measurement.

In order to constructively assess the ten headline KPIs it is necessary to identify the primary objectives of each particular project indicator. (The points noted in the following section relate to *Figure 2.2*.)

1. Project indicator: construction cost

> Defined as a change in the current normalised construction cost of a project at Commit to Construct (point B) compared with one year earlier, expressed as a percentage of the one year earlier cost.
>
> (Report of the Minister of Construction 2000)

This indicator sets out to measure the construction cost of a project this year and compares it to the cost of a similar project constructed last year. 'Rethinking Construction' set a target to reduce the year-on-year cost by 10 per cent. The method of measuring the construction cost involves considering 'two identical structures', when each are:

> completed in successive years and the second is finished for 10% less than the first, then the cost indicator would be – 10%.
>
> (Cook 1999)

Very rarely are two construction projects the same, so in order to establish the cost of two projects, normalisation factors are used to take into account such variables as location, building materials, process and size. The resulting cost indicator is then plotted on the relevant line graph; the benchmark score is then defined. The benchmark score represents the organisation's performance, with the industry standard performance being 50 per cent.

2. Project indicator: construction time

The construction time can be defined as:

> A change in the current normalised construction time of a project at Commit to Construct (point B) compared with one year earlier, expressed as a percentage of the one year earlier time.
>
> (Report of the Minister of Construction 2000)

It is 'Rethinking Construction's' target to reduce the construction time by 10 per cent year on year. Cook (1999) opines that construction time can be measured by comparing two similar projects that are finished a year apart. If the second project is completed in 10 per cent less time after the normalisation factors have been applied, then the time indicator is − 10 per cent.

3. Project indicator: cost predictability

A problem identified by the 'Rethinking Construction' report was the relative uncertainty of the construction costs. This indicator is defined as the

> Change between the actual construction cost at Available for Use (point C) and the estimated construction cost at Commit to Construct (point B).
> (Report of the Minister of Construction 2000)

'Rethinking Construction' set the challenge to make an improvement of 20 per cent year on year.

4. Project indicator: time predictability

The Time predictability indicator has the same reasoning as the Cost predictability indicator as there was a need at the time of production of 'Rethinking Construction' to improve the predictability of time during all stages of the project. The Time predictability is the

> Change between the actual construction time at Available for Use (point C) and the estimated construction time at Commit to Construct (point B), expressed as a percentage of the estimated construction time at Commit to Construct (point B).
> (Report of the Minister of Construction 2000)

5. Project indicators: defects

Cook (1999) defines the defects indicator as how the handover of the project was affected by defects. Four individual terms may be used in order to classify the defects:

- Defect free
- Few defects and available for use on handover
- One or more defects and slight delay
- Major defects that substantially delayed handover

6–7. Project indicators: products and service satisfaction

These two indicators measure the client's satisfaction with the finished product and how satisfied they are with the services they received from the project team. These indicators are measured by asking the client at the end of the project to give the organisation a rating between 0 and 10 for each performance indicator, with 10 indicating totally satisfied.

8. Company indicator: profitability

Cook (1999) defines this indicator as a 'company's pre-tax profit as a percentage of sales'; this indicator has a target to increase the company's turnover and profit by 10 per cent year on year.

9. Company indicator: productivity

The productivity indicator sets out to establish how much turnover each full-time employee generates for the employing organisation. It is 'Rethinking Construction''s target to improve a company's productivity by 10 per cent each year.

10. Company indicator: safety

This indicator strives to measure the number of accidents per 100,000 employees and reduce them by 205 every year.

There are seven basic steps to be followed by construction companies when implementing the ten KPIs. These steps are presented in *Figure 2.3*.

Figure 2.3 establishes that the KPIs have been designed to continually monitor and review the measures of a construction company's performance. There does exist in the construction industry a view that a lack of appropriate performance measurement does hinder organisational and industry improvement. For continuous improvement to occur it is necessary to have performance measures which check and monitor actual performance, to verify changes and the effect of improvement actions. The EFQM EM described in Chapter 4 provides an excellent model for addressing this critical issue.

The following is a generic process and can be adopted/adapted to the other KPI schemes.

Step 1 Decide what to measure

The construction industry has a multiplicity of KPIs and these can be used to measure the performance of a company or an individual project. There is a danger that people involved in the setting and interpreting of KPIs may be utilising incorrect KPIs, resulting in the production of invalid and misleading data. It is vital that a company clearly determines those areas which need to be measured and this must relate to a strategic approach to improvement activities. The selection of appropriate measurement systems and procedures is a very critical activity when undertaking a corporate or

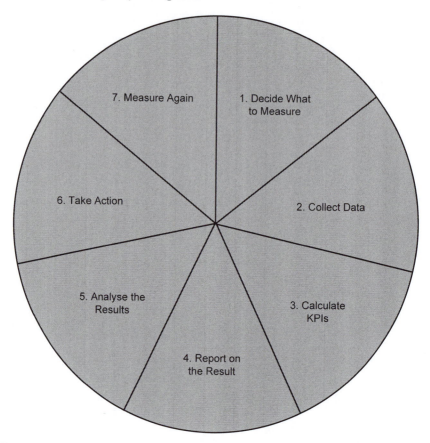

Figure 2.3 The seven stages of implementing KPIs

project-monitoring system. Companies undertake monitoring activities in order to control and evaluate variations, and improvements. A construction firm should ensure that it measures what is important to both the firm and its stakeholders.

In the early stages of KPI deployment companies have a tendency to try and measure too many KPIs; this can result in confusion. The best strategy for a firm is to identify those KPIs that are critical to its activities and commence with those. A good place to start measuring KPIs is to measure the ten standard KPIs, in order to assess the current state of the organisation. It should be noted that the more traditional criteria usually measured in projects, for example costs and schedule times, are not necessarily appropriate for engagement in continuous improvement activities (see Chapter 6). This is because they are not completely effective in identifying the root causes of productivity and quality-problematic issues. Also they do not provide an adequate vision of the potential for improvement and information obtained usually arrives too late to take timely effective corrective actions.

Step 2 Appropriate data collection

Data that needs to be collected will initially come from the contractor's existing records, clients and suppliers. As an example, the information required to apply the Health and Safety KPIs needs to be collected from internal records. Whereas the information required for addressing client satisfaction, in terms of product and service levels, needs to come from the client. In order to be able to accurately measure performance, the information should be valid, reliable and timely.

Step 3 Calculation of KPIs

In order to commence the calculation of KPIs a construction company must first decide which set of data it is going to benchmark against. Many construction firms compare themselves against the 'All Construction' data set. This data set contains all key sectors of the construction industry 'except material suppliers'; however, it may be better for firms to compare themselves against more appropriate KPIs.

Having established a valid methodology for comparing the results, the next stage is to calculate the individual KPIs and a suitable method would be to follow the rules as defined by the KPI Working Group. The calculated scores can then be plotted onto each individual indicator graph, in order to determine the benchmark score. The benchmark score will be a value between 0 and 100 per cent; a value of 50 represents the average performance of other sector companies.

Step 4 Reporting the results of the analysis

Resulting individual benchmark scores have to be both fed back and fed forward if any benefit is to be achieved. The benchmark scores can be plotted onto a radar chart.

Pearson (2002) argues: ' ... the purpose of Key Performance Indicators is to show how a company's performance compares with the average achieved by the industry – shown by the 50% circle in the radar charts.'

Step 5 Analysis of results and implementation

Construction companies must ensure that the resulting analysis is presented in a manner that can be readily understood by those people who are charged with addressing any variances, in the drive for improvements. Too many construction companies stop at the analysis of the results stage, and never obtain the full benefit of using their performance data, in making continuous corporate and project improvements. Fitting actions should be taken in order to maintain corporate strengths and eliminate/reduce corporate and project weaknesses. KPIs cannot by themselves inform a company of what actions need to be taken in order to improve corporate and project performance.

Step 6 Take action

One must remember that the taking of any action has to be made in a timely manner. It is not possible to have effective retrospective corrective actions. Also the actions taken must relate to the issues under consideration. So it is important to correctly establish the cause-and-effect relationship. Having taken some corrective actions it is then important to gauge their effectiveness.

Step 7 Re-measure

Continuous improvement is a cyclical activity; thus, once the first cycle of measuring KPIs has been undertaken and measures put in place to improve the performance, the noted activities should be measured again. This will ensure that the measures that have been taken actually led to an improvement on the previous results (see Chapter 6).

The utilisation of KPIs is based on the strategic action of trying to improve and/or gain a sustainable competitive advantage. It is worth remembering that a construction firm's competitive advantage is based on the value it is able to create for its clients that exceeds the firm's cost of creating it. KPIs can assist in increasing value without increasing corresponding costs for the company and its internal and external stakeholders. When deploying KPIs construction firms should ensure that KPIs relate to their adopted business strategy. A simple checklist is to ensure that:

- Companies focus on KPIs that have an immediate benefit to their business activities and clients.
- Use KPIs that are valid, easily measured and readily understood.
- People get used to reporting and displaying the resulting data.
- KPIs become an integral part of any corporate and project performance assessment activity.
- A company develops a continuous improvement culture.

Benchmarking

Put simply, benchmarking is a form of individual and organisational learning, though such learning has been described as 'Adaptive Learning' (see Chapter 6). Adaptive learning is more related to 'Single Loop Learning', with the inference being that it is more akin to copying than actual learning. However, it can be seen to contribute to the continuous striving for improvement in personal, project and corporate performance. Many construction organisations do now encourage and engage in benchmarking activities as a fundamental business practice, some with the aim of becoming 'best in class'.

Defining benchmarking

Many definitions of benchmarking exist, but it is fundamentally concerned with making valid comparisons between other organisations or projects, and then learning the lessons that these comparisons reveal.

The Royal Mail have defined benchmarking as a structured process of learning from others, internally or externally, who are leaders in a field or with whom legitimate comparisons can be made. The American Productivity and Quality Center define it as a process of continuously comparing and measuring an organisation against other organisations anywhere in the world to gain information on philosophies, policies, practices and measures which will help the organisation take action to improve its performance (APQC).

Benchmarking should not just focus on obtaining performance measures; to become truly effective it has to become part of a construction company's core business strategy, in a drive to keep an organisation at the competitive edge. The essential elements of an effective and efficient benchmarking activity are that the practice is:

- Continuous: benchmarking should not be treated as a 'one-off' exercise; it should be incorporated into the regular planning cycle of the construction organisation, and part of the management of key processes. Its true value is in it being part of an iterative continuous improvement process.
- Systematic: it is important for construction firms to ensure that a valid and consistent methodology is adopted, and that it is actually followed. It is also very important that processes are in place to ensure that good practice is shared across an organisation if it is to obtain the true benefits of engagement with the benchmarking activity.
- Implemented: benchmarking assists in the identification of any gaps, that may exist between an organisation's current performance and that required when performing at 'Best Practice' level; this is achieved by conducting a comparative analysis. A host company must know how 'Best Practice' performance can be achieved, and have the resources to deploy any necessary actions. For improvements to occur at corporate and project levels a clear set of actions need to be established and implemented.
- Good Practice: the identification of absolute 'Best Practice' is not essential for benchmarking to be deemed successful. It is more likely that obtaining a measure of 'Good Practice' would be acceptable.

The European Foundation for Quality Management (EFQM) considers the basic philosophy of benchmarking to be:

- Knowing what you want to improve/learn about (Scoping)
- Identifying the 'Good Practices' in those areas
- Learning from the 'Good Practices' – organisations:
  What they are achieving
  How they are achieving it
- Adapting the Key Insights and incorporating the learning into your own processes.

(Chapter 4 provides further detailed information on this topic.)

In summary, it may be stated that benchmarking empowers construction companies to adopt/adapt and improve organisational practises. The engagement of benchmarking requires a significant focus to be placed on 'process thinking'. This requires an organisation to realise that it consists of a set of processes that are cross-cutting. A process can be defined as a sequence of activities that adds value by producing required outputs from a variety of inputs.

Table 2.1 provides a distinction between the different terms used in the context of Process Management and Benchmarking, with *Figure 2.4* providing a pictorial representation.

Table 2.1 Three types of measure for process

Terminology			Explanation
1. Process	or	Efficiency	Resources consumed in the process relative to minimum possible levels that could be obtained
2. Output	or	Effectiveness	Ability of a process to deliver products or services according to specified specifications
3. Outcome	or	Product/service effectiveness and customer satisfaction	Ability of outputs to satisfy the needs of clients

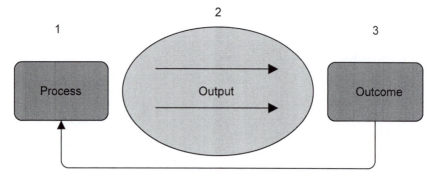

Figure 2.4 Process model related to Table 2.1

Benchmarking provides a means for construction organisations to gauge how well their processes are performing, relative to other internal and/or external organisations, performing similar activities.

Benchmarking can also prove very useful to construction firms when they are trying to:

- obtain an objective assessment of their process(es)' strengths and weaknesses;
- finding ways to stimulate people and groups to engage in improvements;
- overcome internal and external resistance to identified necessary change activities;
- validate (or not, as the case may be) the methods, operations or resources currently utilised.

Benchmarking also assists construction organisations in answering the 'How do you know?' type of questions, such as:

- how do you know that you are achieving a superior performance?
- how do you know your improvement plan will improve corporate and project performance?
- how do you know that you have the best processes available?

The Public Sector Benchmarking Service identifies the following seven main approaches to benchmarking:

Strategic benchmarking

This is used where construction-related organisations are seeking to improve their overall corporate and project performance, by focusing on specific strategies or processes. The key driver for benchmarking is the continuous enhancement of the firm in meeting its strategic aims. And benchmarking is implemented within the context of the development of these core business strategies. Benchmarking is usually undertaken against known examples of best practise, related to the set industrial context.

Performance or competitive benchmarking

This is a process whereby construction firms use performance measures in order to be able to compare their results against those of other similar companies, or processes. This is a common practice in most industrial sectors; measures may include cost per unit of production or profit produced per employee. Benchmarking using this methodology can also be applied within a construction organisation by comparing the performance of individual projects or teams.

Process benchmarking

This approach focuses on specific utilised processes or operations; for example, in construction it could relate to the process of materials handling, with a view to determining and deploying improvements.

Functional and generic benchmarking

Functional or generic benchmarking involves partnerships of organisations drawn from different sectors, all of whom have a desire to improve some specific activity or process. The EFQM has encouraged groups of organisations to work together to benchmark approaches to strategic activities such as knowledge management and process management. (See Chapter 4 for further information.)

External benchmarking

This form of benchmarking can enable the comparison of a construction organisation's functions and key processes; they are compared against good practice in other organisations. The motivator is usually a search for improvement opportunities in business processes.

Internal good practice benchmarking

This is achieved by establishing good practice organisation-wide, usually through the comparison of internal activities or operations. The initiator is usually the sharing of good practice in cross-cutting activities; for example, by carrying out process improvement. This can be done in the context of business planning, which enables the prioritising of specific process improvement projects, allowing results to be compared across business units or projects, in order to identify internal 'best practice', which is then shared.

International benchmarking

Benchmarking can be undertaken internationally as well as nationally.

This is a practice that seeks to identify opportunities for improvement by making comparisons of product/processes or services relating to cost, quality, time, service level and any other key features required by clients.

The relationships between these different types of benchmarking is shown in *Figure 2.5*.

A valid approach to benchmarking would be to select from an appropriate mix of all of the noted methods, and organisational learning is best done when it is carried out within a spirit of partnership and collaboration that enables all parties to learn from each other.

The relationships illustrate that benchmarking can become a central strategy for strategic direction setting and business planning. Benchmarking then

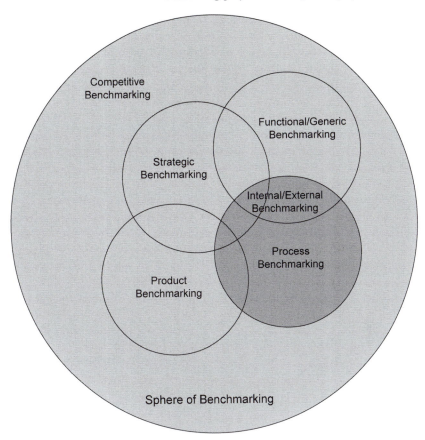

Figure 2.5 Relationships between the different types of benchmarking

becomes a potential improvement strategy for the development of improvement strategies for any key process of the construction organisation.

Benchmarking focuses attention on what is important for the host company. It further provides an approach for creating improved value and quality, but also for improving resource management and productivity. It strikes a balance between stability and renewal and provides a mechanism for studying other organisations to see how they have developed and managed their processes and functions successfully.

Companies that practise benchmarking tend to be more proactive and have a tendency to be externally focused and close to their stakeholders. They are better able to achieve significant improvements in corporate performance and operational competitiveness.

When construction companies embrace the concept of organisational learning, managers and other employees learn and apply new techniques. There is also the opportunity to codify successful behaviour to create new areas of personal competence. Benchmarking can be used as a means to overcome any existing resistance to new ideas, which are often due to staff complacency. Benchmarking can introduce learning as an important component of corporate and project performance. (See Chapter 6 for further information on organisational learning.)

Inter-firm comparison

Benchmarking can be used to critically analyse any aspect of a construction organisation's project or corporate performance; it is also appropriate for both manufacturing and service-industry sectors. Typical organisational functions that are suitable for benchmarking include purchasing, materials control, customer service satisfaction levels and marketing performance, and many more.

For over four decades the Centre for Interfirm Comparison has carried out benchmarking projects and has worked successfully in most sectors of the economy and many countries throughout the world. They are the world leaders in competition benchmarking, sometimes called inter-firm comparison.

Inter-firm comparison is a mutually beneficial activity based on the provision of detailed information, provided in confidence by participating companies, on a comparable basis. Companies decide exactly what information is to be included and exactly how each item is to be defined, and identify any differences in company practices that should be taken into account. Therefore the resulting data is valid and can be used with confidence when making comparisons. Typically about a hundred different ratios are utilised and are shown in full for each participating company anonymously, via the application of coded letters. By comparing their ratios with those of other similar companies it is possible for them to assess their performance against specific set standards. This process can then inform the feedback and feed forward of relevant information leading to continuous project and corporate improvement.

The Centre's inter-firm comparisons are of real practical value to construction firms, at both project and corporate levels. However, firms must ask the correct questions, select the appropriate ratios, critically analyse the results and take timely necessary resulting actions.

Examples of financial benchmarking for construction companies

Performance ratios

$$\text{Profit margin (\%)} = \frac{\text{Profit before Tax}}{\text{Sales}} \times 100$$

This ratio is probably the most useful for a construction business, as it enables a firm to see the bottom line for its activities. A construction business will always be interested in how much profit it is making, in comparison to its sales (turnover), and for this reason the profit margin is often expressed as a percentage of the turnover figure. It is not a sensible strategy for a construction company to have a large turnover figure if it is not generating an acceptable profit margin.

$$\text{Return on Shareholder Funds (\%)} = \frac{\text{New Profit before Tax}}{\text{Total Capital Employed}} \times 100$$

This ratio is obviously important because it establishes for a shareholder or a potential shareholder what they will obtain from their investment in that specific company. Because of the associated risk of investment involved, shareholders will expect a greater return than that which is available from a bank or other financial institution.

$$\text{Credit Turnover} = \frac{\text{Sales}}{\text{Creditors}}$$

This ratio provides a means for a construction firm to assess how its turnover relates to how much it owes its creditors. A construction company will normally only want to pay its creditors once it has received payment from its debtors. Thus the figures for Creditor Turnover and Debtor Turnover should stay approximately the same.

$$\text{Debtor Turnover} = \frac{\text{Sales}}{\text{Creditors}}$$

See explanation for creditor turnover.

$$\text{Stock Turnover} = \frac{\text{Turnover}}{\text{Stock}}$$

This ratio allows the firm to see clearly how the quantity of materials they have in stock compares to their turnover figure. A firm will aim to have as little stock as is practicable, in order to achieve the greatest turnover.

$$\text{Debtor Collection (days)} = \frac{\text{Debtors}}{\text{Sales Turnover}} \times 365$$

This ratio allows a company to calculate how many days on average it is taking to recoup its debts.

$$\text{Creditors Payment (days)} = \frac{\text{Trade Creditors}}{\text{Purchasers}} \times 365$$

This ratio allows a company to calculate how many days it should restrain before paying its creditors, and a useful rule of thumb is that these two ratios should be roughly the same.

$$\text{Net Asset Turnover} = \frac{\text{Sales}}{\text{Net Sales}}$$

This ratio allows a company to assess how much its net assets are worth in relation to its turnover figure. A company does not want to have too much of its value in its fixed assets. If this is the case, it may want to consider other arrangements, such as rental of such assets or the hiring of its plant and equipment.

$$\text{Return of Equity (\%)} = \frac{\text{Profit after Tax}}{\text{Shareholders Funds (capital + reserves)}} \times 100$$

This ratio is quite similar to the ratio for return on shareholder funds. It allows the company to assess whether or not it is achieving sufficient profit, based on the amount of money invested by shareholders.

$$\text{Return of Net Assets (\%)} = \frac{\text{Profit before Tax}}{\text{Net Assets}} \times 100$$

This ratio shows how much gross profit a company is achieving in relation to the value of its net assets.

$$\text{Current Ratio} = \frac{\text{Current Assets}}{\text{Current Liabilities}}$$

This ratio is critical because it predicts the solvency of the company. Bankers, auditors and accountants, amongst others, will be very interested in the results of this ratio. A construction firm's liabilities should not exceed its assets, as this is not beneficial for the company in the long term.

$$\text{Liquidity Ratio} = \frac{\text{Current Assets less Stocks}}{\text{Current Liabilities}}$$

This ratio is similar to the current ratio, but it provides for an even more stringent test. Banks refer to it as the 'acid test' ratio. It deals mainly with 'liquid' elements, e.g. money. Usually things such as overdrafts would not be included because they are not such an urgent debt as are creditors.

The results of these financial ratios can thus be compared with similar construction organisations in order to obtain valuable Industrial Comparable qualitative data, to be incorporated into its decision-making process.

Financial ratios example

Corporate financial fitness depends very much upon healthy financial proportions, rather than on absolute amounts. Students like to be told, for example, that current assets/current liabilities should be in the ratio of 2:1. However, company circumstances vary; for example, the stock turnover in a supermarket will clearly be well above that of a construction contractor. Thus it is vital that any inter-firm comparison be based on similar types of companies.

Also, health often depends on the actual items making up the ratio, so that a poor current assets/current liabilities ratio may be satisfactory, if the current assets are wholly cash balances.

The above means that good balance-sheet analysis requires a combination of common sense and sound financial judgement.

Table 2.3 provides an example of the ratios applied to a balance sheet.

Table 2.2 Balance sheet example data

Condensed income statements for the year ending 31 December

	A £000s	B £000s
Sales	2,000	2,400
Cost of Sales	1,200	1,400
Gross Margin	800	1,000
Expenses	400	440
Income before taxation	400	560
Less taxation	170	252
Income after taxation	230	308
Dividends paid	160	160
Income retained	70	148

Condensed Balance Sheet as of 31 December

CAPITAL EMPLOYED	A £000s	B £000s	Fixed Assets	A £000s	B £000s
Share capital – issued and fully paid ordinary £1 share	800	800	Net of depreciation		
Retained earnings	200	348	Plant/equipment	600	800
9% debenture	100	100	Motor vehicles	200	300
	1,100	1,248		800	1,100
Current liabilities			Current assets		
Creditors	270	300	Inventory	400	540
Taxation	170	252	Debtors	180	200
Dividends Proposed	160	160	Bank	320	120
	1,700	1,960		1,700	1,960

Table 2.3 Ratio analysis applied to the balance sheet data

Time Period A	Time Period B

Profitability Ratios

$$\frac{\text{Profit}}{\text{Capital Employed}} \times 100$$

$\dfrac{400}{1,100} \times 100 = 36.36\%$	$\dfrac{560}{1,248} \times 100 = 44.87\%$
	An increase of 8.5%

$$\frac{\text{Profit}}{\text{Sales}} \times 100$$

$\dfrac{400}{2,000} \times 100 = 20\%$	$\dfrac{500}{2,400} \times 100 = 23.3\%$
	An increase of 3.3%

$$\frac{\text{Sales}}{\text{Capital Employed}}$$

$\dfrac{2,000}{1,100} = 1.8$ Times	$\dfrac{2,400}{1,248} = 1.9$ Times
	An increase of 0.1

Note
Overall profitability shows an increase

Liquidity Ratios

$$\frac{\text{Current Assets}}{\text{Current Liabilities}}$$

$\dfrac{900}{600} = 1.5 : 1$	$\dfrac{860}{712} = 1.2 : 1$
	A fall in long-term liquidity of 30p in the £

$$\frac{\text{Quick Assets}}{\text{Current Liabilities}}$$

$\dfrac{500}{600} = 0.83 : 1$	$\dfrac{320}{712} = 4.45 : 1$
Note	*Short-term liquidity has fallen by 38p in the £*
QA taken as debtor's + bank cash	

Table 2.3 (continued)

Time Period A	Time Period B

$$\frac{\text{Debtors}}{\text{Sales}} \times 365$$

$$\frac{180}{2,000} \times 365 = 32.85 \text{ days}$$

$$\frac{200}{2,400} \times 365 = 30.40 \text{ days}$$

A reduction of 2.45 days

$$\frac{\text{Sales}}{\text{Stock}}$$

$$\frac{2,000}{400} = 5 \text{ Times}$$

$$\frac{2,400}{540} = 4.4 \text{ Times}$$

Increase in stock held

General statement:

- Liquidity situation is moving from bad to worse.
- The value of such analysis is improved by comparing and contrasting different financial trading periods.

Questions for the reader

Here follows a number of questions related specifically to the information presented within this chapter. Try to attempt each question without reference to the chapter in order to assess how much you have learned. The answers are provided at the end of the book.

Case study

From the data provided below, Mr. Smith, the Managing Director of Smith's PLC, has asked you to explain the current financial situation of his company. He has requested that you use as the basis of your analysis the KPIs (financial ratios) of

Liquidity ratios

$$\frac{\text{Current Assets}}{\text{Current Liabilites}} = \text{Current Ratios}$$

A measure of the level of safety involved in relying on current assets being sufficient to pay current liabilities (2:1).

$$\frac{\text{Quick Assets}}{\text{Current Liabilites}} = \text{Acid Test}$$

A measure of the level of safety involved in relying on 'quick' assets to cover liabilities (1:1).

(Note: 'quick' assets are usually taken as cash and debtors.)

$$\frac{\text{Debtors}}{\text{Sales}} \times 365 = \text{Average time taken to obtain payment from debtors}$$

$$\frac{\text{Sales}}{\text{Stock}} = \text{The number of times stock is turned over per trading period}$$

Balance sheet ratios

$$\frac{\text{Profit}}{\text{Capital Employed}} \times 100 = \text{Return on capital}$$

Profit is usually taken before tax. This ratio measures the creation of wealth relative to the economic wealth tied up in the process.

$$\frac{\text{Profit}}{\text{Sales}} \times 100 = \text{amount of profit generated by sales made during trading periods}$$

$$\frac{\text{Sales}}{\text{Capital Employed}} \times 100 = \text{the amount of times that the capital is turned over}$$

Mr. Smith would like your comments to specifically address liquidity and profitability.

Case study: balance sheet data

The following are the summarised balance sheets at 30 April 2009–10 for A. Smith's PLC.

Balance Sheets as at 30 April

	2009 £	2010 £
Issued Capital		
Ordinary Shares	450,000	450,000
Revenue Resources	357,000	237,000

	2009 £	2010 £
11% Debentures	300,000	300,000
	CE 1,107,000	987,000
Current Liabilities		
Creditors	QA 673,000	641,000
Overdraft (secured)	266,200	432,000
Taxation	33,800	–
	CL 973,000	1,073,000
	2,808,000	2,060,000

Balance Sheets as at 30 April

	2009 £	2010 £
Fixed Assets		
Freehold Property (at cost, 1992)	336,000	336,000
Plant & Equipment (less depreciation)	410,000	260,000
	746,000	596,000
Current Assets		
Stocks	182,000	90,000
Work in Progress (jobbing work)	132,200	123,000
Uncompleted Contracts (less progress payments)	480,000	580,000
Debtors	540,000	663,000
	CA 1,334,000	1,464,000
	2,080,000	2,060,000

The following information was obtained from the trading and profit-and-loss accounts at 30 April

	2009 £	2010 £
Completed Contracts	2,400,000	2,130,000
Completed Jobbing Work	100,000	100,000
	Sales 2,500,000	2,230,000
Less Cost of Sales	1,945,000	1,875,000
Plant & Equipment (less depreciation)	410,000	260,000
	746,000	596,000
Gross Profit	555,000	355,000
Less General Expenses (incl. depreciation)	385,000	375,000
Net Profit/Loss before Tax	Profit 170,200	(20,000)
Taxation for Year	51,000	–
Net Profit/Loss after Tax – Retained	119,000	(20,000)

Further reading

McCabe, S. (2001). *Benchmarking in Construction*. Oxford: Blackwell Science.
McGeorge, D. and Palmer, A. S. (1997). *Construction Management New Directions*. Oxford: Blackwell Science.

Phillips, R. Robert and Freeman, Edward (2003). *Stakeholder Theory and Organisational Ethics*. Berrett-Koehler Publishers.

References

Cook, A. (1999). How do You Measure Up? *Building*, pp. 24–30.
Freeman, R. Edward (1984). *Strategic Management: A Stakeholder Approach*. Boston: Pitman.
Great Britain, The KPI Working Group (2000). *KPI Report for the Ministers for Construction*. London: Department of the Environment, Transport and the Regions.
MacDonald, J. and Turner, S. (1996). *Understanding Benchmarking in a Week*. Hodder & Stoughton Educational, London.
Mitchell, R. K., Agle, B. R. and Wood, D. J. (1997). 'Toward a Theory of Stakeholder Identification and Salience: Defining the Principle of Who and What Really Counts' (http//www.jstor.org/stable/259247). *Academy of Management Review* 22(4): pp. 853–86. 10.2307 / 259247 (http://dx.doi.org/10.2307%2F259247). http://www.jstor.org/stable/259247.
Pearson, A. (2000). Don't go KPI nuts. *Building*, pp. 44–45, UK.

3   ISO 9001:2008

This chapter fully explains the philosophy and concepts of quality assurance built upon the foundations of the certificated process model of ISO 9001:2008. The constituent parts of the model are presented and explored for the reader. The advantages of certification are provided, and the important issue of how to demonstrate compliance with the standard is discussed. A flow diagram designed by the book's authors and incorporated within the text should prove to be a very useful tool for construction companies engaging in the implementational process.

Learning outcomes

By the end of this chapter the reader will be able to demonstrate an understanding of:

- The fundamental principles of quality assurance
- The process model approach to quality assurance
- The key issues to be addressed when implementing ISO 9001:2008
- The benefits of ISO 9001:2008 deployment for construction-related organisations.

Introduction

This chapter provides an introduction to a certificated quality standard which has been deployed by many thousands of service and manufacturing providers, including construction companies. It further offers a rationale for organisations to fully address the important issue of quality. The concept of product and service quality is an issue which is given great consideration by the purchasers of products and/or services, and thus must be given equal consideration by product and service providers.

The ISO 9001:2008 is an international standard on quality management systems (QMS). It enables the application of a uniform approach to quality management by construction organisations of all sizes, within the provision of a uniform model.

The International Organisation for Standardisation (ISO) first published this international standard in 1987, and it was revised in both 1994 and 2000.

However, in 2008 it underwent a further revision; this latest revision was undertaken in order to ensure that it fully satisfied the requirements of service and manufacturing organisations. Thus it is now referred to as ISO 9001:2008; it should be noted that within the context of ISO 9001:2008 the term product applies to both tangible and intangible products.

A valid strategy for construction organisations seeking to gain a sustainable competitive advantage is that of differentiating its product or service. This differentiation can be achieved by the provision of a quality focus. If the providers of products or services are to be truly competitive then the quality of their products and/or services must match customer expectations. This chapter explores how ISO 9001:2008 can be utilised by product- and service-providing organisations, in order to obtain a highly valued sustainable competitive advantage. The ISO 9001:2008 standard, along with its associated constituent parts and advocated advantages of implementation, are explained within the text.

Also within the chapter the problematic issues of application are explored, and solutions advocated; thus the reader will gain a full appreciation of how to implement the quality standard.

Corporate planning and ISO 9001:2008

Most construction organisations engage in some form of corporate planning activity, although some would not recognise it as such. Every small and medium sized company (SME) that has decided that it will increase turnover, improve return on investment and increase market share has engaged in corporate planning activity. In its more sophisticated form, a corporate plan will involve a team of experts working closely with the senior management of a large company, applying a range of advanced computer modelling techniques designed to consider future probabilities and designing an appropriate strategic approach.

Successful corporate planning is very much dependent upon the provision of accurate and valid information. It requires frequent testing of the feasibility of proposals and constant input from managers at all levels of the construction company.

Corporate plans and manpower plans must be mutually interdependent and set within a truly holistic context, and this has two main facets.

- Corporate policy shaping manpower plans: If corporate policy determines that a company is to expand and diversify then new employees and skills/competences will probably be required to make this possible. This is an important fact when considering the specific requirements of ISO 9001:2008. Different parts of a business and different functions should be considered separately when examining their manpower implications, but a holistic overview must be taken. This will help to prevent the regrettably all too frequent occurrences where one part of a business is declaring redundancies, while in another part a desperate recruitment drive is being undertaken. If such situations are identified in time, corrective action can be taken through retraining and redeployment to avoid the worse consequences of both.

- Manpower plans influencing corporate policy: Manpower resources can act as a very real constraint upon the achievement of specific corporate objectives and the implementation and continual application of ISO 9001:2008. Manpower is a valuable resource and as such it must be available in the correct proportion, along with the other Four Ms (materials, money, machines and management).

When an organisation is assured that it's making the optimum use of its existing manpower and has identified what possible changes in demand for manpower corporate policy may cause (e.g. the deployment of ISO 9001:2008), it is ready to be proactive. But the identification of manpower needs is not a once-and-for-all exercise, it should be a dynamic activity. A whole range of other factors may influence the demand for a particular number or types of employees.

These factors include:

- Possible market fluctuations affecting demand for a construction firm's products and/or services and hence the number of employees required to make and/or provide them.
- Changes in the availability of raw materials, affecting levels of production and hence organisational manpower requirements.
- Technological advances which preclude the need for some jobs and possibly change the skills/competences required to perform others. For example, a decision to computerise certain systems; this can usually be considered well in advance of the event, preventing panic reactions. Other changes may require a more immediate response.
- Government intervention (in health and safety for example) may lead directly or indirectly to the creation of new jobs or the realignment of some responsibility.
- Mergers and takeovers can affect every aspect of corporate life and objectives are likely to change, as may the culture of the whole organisation.
- Internal problems such as unexpected industrial relations difficulties.
- Changes in the cost of labour relative to that of other resources.

Each of the above factors forms a variable in a company's manpower demand planning and management task. It must be remembered that ISO 9001:2008 will have some impact upon the process utilised by organisations and hence a corresponding impact upon a construction organisation's manpower requirements.

Senior management may set the policy and objectives for the company but they should not forget that the implementation issues involve everyone in the construction organisation, especially the quality issues. This is a very important point when endeavouring to deploy ISO 9001:2008 successfully.

People within construction organisations are the mechanism by which implementation takes place. The extent to which people within construction organisations influence objectives and deployment has been identified by Miles and Snow (1983) who postulated that "Organisations [including construction organisations] do not have objectives but people have values." Thus organisational personnel may ultimately have an impact upon corporate objectives. With regard to construction organisational personnel it is safe to state that it is really about having the correct number of competent staff required to perform set project and corporate tasks. However, we should not forget that "Labour productivity is a measure of how efficiently the human resources are being used. To some extent it combines an assessment of both efficiency and effectiveness since poor allocation of people to jobs (effectiveness) would result in low productivity" (Johnson and Scholes 1984).

Therefore having the correct staff with appropriate competences and skills can ultimately impact upon a construction organisation's ability to deliver the required quality specification of product and/or service required by its customers.

Quality assurance in construction

Essentially there is a requirement to provide an assurance that design and construction aspects have the capability to produce a product that is effective and economic, whether that product is the design of the building or the construction of the building, and these aspects are not mutually exclusive. The pursuit of quality commences with the client and continues through the production process to the utilisation of the building. Quality assurance is therefore an integral part of the "total building process".

Quality is in many ways a subjective entity and to a certain extent is a matter of personal judgement. To provide a clearer view of the meaning of quality, Griffith (1990) defines a number of aspects which should be considered:

- Function: does the building meet the requirement?
- Life: is the building durable?
- Economy: does the building represent value for money?
- Aesthetics: is the building pleasing in appearance and compatible with its surroundings?
- Depreciation: is the building an investment?

The interpretation and measurement of quality are as ambiguous as its perception. Clients will have their own ideas of the quality required to meet stated needs and desires.

The architect aims to provide value for money, assisted by the quantity surveyor; all are bounded by providing an acceptable standard of construction. On site, the quality of work is dependant on the skills and the application of the craft operatives or 'workmanship'. It can therefore be hypothesised that one's view of quality is dependent upon one's involvement and role in the overall construction process.

Quality in construction can therefore be determined by expectation. Dayton (1988) emphasises that ' ... management of quality and quality itself are closely related to a number of various expectations surrounding the performance of buildings, these being quality, durability and reliability'.

Quality assurance systems

Quality assurance is concerned with planning and developing the technical and managerial competence to achieve the desired objectives, whether these are set by the client or the host company. Quality assurance is also concerned with the management of people, addressing the roles, duties and responsibilities of individuals within the construction organisation. Quality assurance is primarily the responsibility of management, but its structure and implementation must become part of the total holistic construction organisational framework, and, as previously noted, related to the corporate personnel strategy.

Quality assurance must also be an important aspect of the marketing and promotional strategy of the construction firm. Only when quality assurance pervades the entire construction organisation and becomes an integral and recognised aspect of its operations will it foster the potential to become truly successful in providing an organisational competitive advantage.

Quality assurance must be actively employed throughout the total building process, from initial briefing and conceptual design, through the assembly process, to the completion of the project. It is essential that clear communication is planned for and encouraged, in particular at the critical interfaces of project responsibility and control.

Quality assurance application

Quality assurance is concerned with developing a 'formal' structure, organisation and operational procedures designed to ensure the specified quality is attained throughout the total building process. The construction industry can be divided into five broad sectors where quality assurance is applicable:

- By the client, in the production of the project brief.
- By the designer, in the design and specification process.
- By the manufacturers, in the supply of materials, products and components.
- By the contractors (and subcontractors), in construction, supervision and management processes.
- By the user, in the utilisation of the new structure.

There are few standards and codes that affect the client and the final procurement and use of the building, with the majority of quality assurance applications being more related to the manufacturing arm of the construction industry.

ISO 9001:2008 implementation provides a framework for the five sectors noted above. Successful implementation provides a system which can be certified by a recognised body. This in turn provides both monitoring and enforcement authority to an independent third party.

Competitive advantage and quality

As noted previously, for a construction organisation to be profitable and sustain growth it must have a sustainable competitive advantage, sometimes referred to as a competitive edge. Johnson and Scholes (1984) identified three valid strategies for organisations to attain a sustainable competitive advantage; these are:

- Least cost
- Focus
- Differentiation

A least-cost strategy

A least-cost strategy is based upon the reduction of fixed and variable overheads per unit of production or service provision. This enables the manufacturers or service providers to offer their goods or services at a lower price per unit than respective competitors. Least cost is usually associated with high-volume production or provision. It is really built upon the foundations of economies of scale.

Focus as a strategy

Focus as a sustainable competitive advantage advocates that an organisation concentrate upon what it is good at, i.e. its distinctive competence. If a construction organisation is good at, for example, small renovation works, it would make little strategic sense to diversify into Civil Engineering. Peters and Waterman (1982) have termed this 'Sticking to the Knitting [and note] ... least successful as a general rule are those companies that diversify'. There are certain advantages here related to learning curve theory.

Differentiation as a strategy

Differentiation implies that a construction organisation's product or service is differentiated in some way from its competitors. It is possible that a product or service can be differentiated via a quality aspect. A product or service may be taken as being synonymous with quality (or your customer's definition/perception of quality). ISO 9001:2008 is very much related to being able to demonstrate that a construction organisation can deliver customer quality requirements.

The three means noted above for obtaining a strategic competitive advantage are not mutually exclusive. It is possible for a construction company to focus upon a particular product or service by which differentiation is obtained via a quality focus and ISO 9001: 2008 could be part of its competitive strategy. If the service chain becomes large enough and/or a large volume of products are manufactured then the noted advantages of least cost may also manifest themselves.

Question one for the reader

The construction industry can be divided into five broad sectors where quality assurance is applicable. Identify these sectors. The answer is provided at the back of the book.

Project roles within a quality assurance system

The following provides an overview of the roles provided by on-site construction staff members in relation to ISO 9001: 2008; however, only a generic approach is possible within the context of this chapter.

Site manager

The site manager carries the burden of accountability for ensuring all site staff under their direction perform their duties in compliance with the project quality assurance system. They are the main link between head office, the client and the project, and as such it is their responsibility to ensure the successful completion of the project, to the complete satisfaction of the client (according to contract conditions and specifications).

The site manager ensures that all work carried out conforms to the specified standards and design criteria. They are charged with enforcing the quality assurance system throughout the project, with due regard being given to the performance of subcontractors, and ensuring conformity to specified standards by all subcontractors and suppliers.

Any discrepancies in information or flaws in design should be notified to the project administrator for their consideration as part of the quality assurance process. All internal and external correspondence, on-site checks and matters arising which might affect the successful completion of the project are to be fully documented and acted upon in accordance with the quality procedures manual.

Any action necessitated out of the quality assurance documentation process must be initiated and progress monitored in the appropriate manner. All quality procedures initiated have to be communicated to the quality manager for inclusion into the company quality assurance system.

Site engineer

The site engineer has responsibility for setting out of the works as directed by the site manager, and must carry out their duties in full compliance with stated specifications and design criteria. They are required to fully document all their site activities in the required manner, keeping detailed records of:

- Site levels;
- Verbal instructions from the client's engineer;
- Discrepancies in information supplied by the client's representative;
- Events necessitating a deviation from current designs or specifications, reporting any nonconformity directly to the site manager.

Quantity surveyor

The main responsibility of the quantity surveyor is the financial control of the project. But during the performance of their measurements for payments they should note any defective workmanship or nonconformity with the required standards or specifications. They need to keep the site manager informed of any such items, discussing possible causes and any remedial action required.

Trade and contract foremen

All trades foremen or subcontractor's foremen have a responsibility to ensure that work carried out by persons under their direction is done to the required standard as specified by the site manager, and with due regard to plans, schedules and specifications, reporting any nonconformity directly to the site manager. All defects or nonconformity identified during the course of their work must be documented in accordance with procedures for submission to the site manager.

Stores foreman

The stores forman's responsibilities include the checking of all invoices and materials for correctness to specifications, when delivered to site. They should also ensure the correct storage of all materials, reporting defects or nonconformities arising from deviations in procedures directly to the site manager. These should also be fully documented.

Site operatives

Site operatives have a responsibility to perform their works to the required standard and as specified by their foreman or site manager; thus they must have the prerequisite skills and competences (again, a key aspect of an organisation's manpower plan and ISO 9001:2008 requirements).

During the execution of their duties they are to report defective materials or contradictions of information to their foreman or the site manager.

Outline of a project quality assurance system

The project quality assurance file details the project particulars for use during the effective management of a construction project. The quality assurance system needs to be documented in a way which makes all project information easily accessible to the user, and ensures consistent use of the quality assurance documentation procedures, for all site functions.

Quality assurance project file overview

Project details

This section on project details lists the general information required for reference purposes by the site management team; this should consist of the following information:

- Details on the client
- Project address
- Brief description of the project
- Commencement date
- Completion date

Contract directory

The contract directory lists all persons or organisations who are involved in the project, detailing their involvement and giving their address, telephone and fax numbers. This information, together with the project details is essential for the effective communication of the management team and the co-ordination of all on-site participants/activities.

Site management structure

This is usually presented as an organisational tree or flow chart, indicating all site staff and showing the lines of authority and autonomy, making reference to any individuals or department at head office that provide a supporting role to the individual site staff.

Subcontractors

This section details all subcontractors on site, providing details of foreman, anticipated labour force and details of any pre-contract undertakings or agreements. For example, specialist lifting equipment or machinery to be supplied by the main contractor. Details of head-office organisation and individuals directly responsible for the subcontractors are also noted.

A subcontractor's programme is included and requires incorporating into the main contractor's programme.

Material specifications

The material specifications section details all material types to be used on the project, noting all details of required specifications, storage and handling details. Also included are any additional considerations, for example lengthy order times or details of especially expensive and fragile items.

Project programme

A detailed project programme will be included, giving full particulars of start and finish dates of all contractors and subcontractors, durations, available float times, minimum and maximum resources and projected valuations.

Quality policy

The quality policy details the construction company's objectives and commitment to quality and states the standard to which the company's quality system conforms, for example ISO 9001:2008. It is imperative that all staff fully understand this document and that it is implemented, as it will form the basis of the third-party audit conducted by the appropriate certification body.

Construction Design and Management regulations (CDM)

The company CDM file must be represented in the project quality assurance file and should include details of the following:

- Method statements
- Plant and equipment identification
- Hazard risk assessment
- Subcontractors' risk assessment and method statements
- COSHH and risk assessment records
- Health and Safety Executive (HSE) notification of project
- Company Health and Safety (H&S) policy
- Company insurers
- Details of emergency procedures

Full documentation of the above will not be contained within the project quality assurance file, but a brief summary of each entry with specific reference to the full documentation of each is required.

Advocacy of quality

From the previously noted three strategic competitive strategies it can be seen that quality issues for both product and service providers are of great importance. A construction company can reduce its price to increase its market share, but that could prove to be a high-risk strategy or tactic. Competitors can quickly follow suit, and not only is the organisation back where it started, but it is likely to be worse off, having started a price war.

The Department of Trade and Industry through the Enterprise Initiative state clearly that

> successful companies [in the 21st century] will become winners through increasing their return on investment (profit) by increasing their market share through consistently demonstrating to customers that they fully satisfy their needs and expectations [as can be demonstrated by ISO 9001:2008] and they will even lead the customers to expect more – which only they can provide.
>
> (Hogg 1990)

The above-noted customer orientation for a construction organisation comes from assessing its customers' needs and expectations – that is quality – and organising and resourcing so it can fully meet them. Successful construction companies will also recognise the need to satisfy those needs and expectations at the first attempt, that is to get it right first time.

Quality does not simply mean the best; a quality product or service is one which gives the customer what they want – a product or service that is fit for purpose.

Quality is more than the quality of a product or service and must cover every activity of the construction organisation. The commitment of every person in the construction company is required and this usually leads to a change in management and workforce's attitude and relationships.

If the above can be obtained then quality improvements are perfectly feasible and deliverable. The specific benefits of a quality focus have been identified by the Plastics and Rubber Institute:

> Benefits are company wide involvement, improved productivity and efficiency, reduced costs and the ability to provide a product that consistently meets customers' needs, so improving customer confidence and service and retaining or increasing market share.
>
> (The Institute of Quality Assurance 1990)

Quality does not however just happen, but requires a professional approach and positive action by senior managers, engineers and technologists in every department of a construction company. Dale (2004) notes that 'What we are talking about here [truly integrating a quality approach] is a long term cultural change.'

Construction companies must do all they can to provide the quality product or service demanded by their clients. Clients need to be assured that the quality provided will be that specified and agreed upon. One means of providing this client assurance is to implement the quality management system of ISO 9001:2008.

ISO 9001:2008

Historical context

As previously noted, the rationale for implementing a quality management system within a construction firm is to provide assurance that the firm can meet client's requirements. Further, it should lead to the attainment of a sustainable competitive advantage. ISO 9001:2008 has been advocated as a strategy for achieving an improvement in the effectiveness, flexibility and competitiveness of construction companies. ISO 9001:2008 aims to assist in producing a superior performance from the whole corporate enterprise. This results in improved quality products and services, delivery and administration, which ultimately satisfy the client's functional and aesthetic requirements within defined cost and completion constraints.

ISO Technical Committee 176 is the body charged with the task of producing and updating quality standards and the ISO 9001:2008 standard replaces the previous ISO 9000 series of standards. The new standard was first published in November 2008, and now organisations applying for certification have to comply with the new standard. At the time those with certification to the old standard were given one year to comply with the new standard.

The ISO 9001:2008 Series has been developed to assist organisations of all types and sizes to implement and operate an effective quality management system. The format consists of:

- ISO 9000 – an introduction to quality management systems and the vocabulary of quality management.
- ISO 9001:2008 – specifies requirements for quality management systems for use where an organisation's capability to provide conforming products needs to be demonstrated.
- ISO 9000:2005 – provides guidance related to quality fundamentals and vocabulary.
- ISO 9004:2000 – provides guidance on the implementation of a broadly based quality management system to achieve continual improvement of business performance.
- ISO 1011 – relates to managing and conducting internal and external quality management system audits.

Together they form a cohesive set of quality management system standards.

The new revisions incorporated into ISO 9001:2008 take into account previous experience with quality management systems from practitioners. They should result in a closer alignment of the quality management system with the needs of the day-to-day running of a construction organisation.

The standards are compatible with the environmental management standards, can be readily applied to small, medium and large organisations in public and private sectors, and equally applicable to users in manufacturing, service and software fields.

Quality management principles (incorporated into ISO 9001:2008)

Managing an organisation successfully requires a systematic approach. Success can result from implementing and maintaining a management system which is designed to continually improve performance by addressing the needs of all interested parties. Eight quality management principles have been identified to facilitate the achievement of project and corporate quality objectives. These are:

- **Customer focused organisation** – organisations depend on their customers and therefore should understand current and future customer needs. They have to meet customer requirements and strive to exceed customer expectations. This approach would result in the following benefits:
  - increased revenue and market share obtained through flexible and fast responses to market opportunities;
  - increased effectiveness in the use of the organisation's resources to enhance customer satisfaction;
  - improved customer loyalty leading to repeat business;
  - researching and understanding customer needs and expectations;
  - ensuring that the objectives of the construction organisation are linked to customer needs and expectations;
  - communicating customer needs and expectations throughout the construction organisation;
  - measuring customer satisfaction and acting on the results;
  - systematically managing customer relationships;
  - ensuring a balanced approach between satisfying customers and other interested parties (such as owners, employees, suppliers, financiers, local communities and society as a whole).

- **Leadership** – leaders establish unity of purpose, direction and the internal environment of the construction organisation. They create the environment in which people can become fully involved in achieving the organisation's objectives, resulting in the following key benefits:
  - people will understand and be motivated towards the construction organisation's goals and objectives;

- activities are evaluated, aligned and implemented in a unified way;
- miscommunication between all levels of an organisation will be minimised, if not eradicated;
- considering the needs of all interested parties including customers, owners, employees, suppliers, financiers, local communities and society as a whole is achieved;
- establishing a clear vision of the organisation's future;
- setting challenging but achievable goals and targets;
- creating and sustaining shared values, fairness and ethical role models at all levels of the organisation;
- establishing trust and eliminating fear;
- providing people with the required resources, training and freedom to act with responsibility and accountability;
- inspiring, encouraging and recognising people's contributions.

- **Involvement of people** – people at all levels are the essence of a construction organisation, and their full involvement enables their abilities to be used for the organisation's benefit and shall empower the attainment of:

  - motivated, committed and involved people within the organisation;
  - innovation and creativity in furthering the construction organisation's objectives;
  - people being accountable for their own performance;
  - people eager to participate in and contribute to continual organisational improvement;
  - people understanding the importance of their contribution and role in the construction organisation;
  - people identifying constraints to their performance;
  - people accepting ownership of problems and their responsibility for solving them;
  - people evaluating their performance against personal goals and objectives;
  - people actively seeking opportunities to enhance their competence, knowledge and experience;
  - people freely sharing knowledge and experience;
  - people openly discussing problems and issues in a drive to learn and improve.

- **Process approach** – a desired result is achieved more efficiently and effectively when related resources and activities are managed as a process, resulting in the following key benefits:

  - lower costs and shorter cycle times, through the effective use of resources;
  - improved, consistent and predictable results;
  - focused and prioritised improvement opportunities;
  - systematically defining the activities necessary to obtain a desired result;

- establishing clear responsibility and accountability for managing key activities;
- analysing and measuring the capability of key activities;
- identifying the interfaces of key activities within and between the functions of the construction organisation;
- focusing on factors such as resources, methods and materials that will improve key activities of the construction organisation;
- evaluating risks, consequences and impacts of activities on customers, suppliers and other interested parties and managing accordingly.

- **System approach to management** – identifying, understanding and managing a system of interrelated processes for a given objective, this contributes to the effectiveness and efficiency of the construction organisation, and enables:

  - integration and alignment of the processes that will best achieve the desired results;
  - an ability to focus efforts on the key processes;
  - providing confidence to interested parties, as to the consistency, effectiveness and efficiency of the construction organisation;
  - structuring a system to achieve the organisation's objectives in the most effective and efficient way;
  - understanding the interdependencies between the processes of the system;
  - structured approaches that harmonise and integrate processes;
  - providing a better understanding of the roles and responsibilities necessary for achieving common objectives and thereby reducing cross-functional barriers;
  - understanding organisational capabilities and establishing resource constraints prior to taking action;
  - targeting and defining how specific activities within a system should operate;
  - continually improving the system through measurement, evaluation and reflection before taking actions.

- **Continual improvement** – continual improvement should be a permanent objective of all construction organisations, and this should provide:

  - a performance advantage through improved organisational capabilities;
  - alignment of improvement activities at all levels to an organisation's strategic intent;
  - a flexibility to react quickly to opportunities;
  - employing a consistent organisation-wide approach to continual improvement of the construction organisation's performance;
  - providing people with training in the methods and tools of learning and continual improvement;
  - making continual improvement of products, processes and systems an objective for every individual in the construction organisation;

- establishing goals to guide, and measures to track, continual improvement;
- recognising and acknowledging improvements.

- **Factual approach to decision making** – effective decisions are based on the logical and intuitive analysis of data and information and incorporating it into an effective decision-making process, thus providing the following benefits:

  - informed and implementable decisions;
  - an increased ability to demonstrate the effectiveness of past decisions through reference to factual records;
  - increased ability to review, challenge and change opinions and decisions;
  - ensuring that data and information are sufficiently accurate and reliable;
  - making data accessible to all those who need it;
  - analysing data and information using valid methodologies;
  - making decisions and taking action based on factual analysis, balanced with experience and intuition.

- **Mutually beneficial supplier relationships** – mutually beneficial relationships between the construction organisation and its suppliers enhance the ability of both organisations to create value, resulting in:

  - flexibility and speed of joint responses to changing market or customer needs and expectations;
  - optimisation of costs and resources;
  - establishing relationships that balance short-term gains with long-term considerations;
  - pooling of expertise and resources with partners;
  - identifying and selecting key suppliers;
  - clear and open communication;
  - sharing information and future plans;
  - establishing joint development and improvement activities;
  - inspiring, encouraging and recognising improvements and achievements by suppliers.

(Adapted from BSI 1999)

ISO 9001:2008 and certification

ISO 9001:2008 is now the only standard against which companies may be certificated. Therefore, when the product or service does not require certain quality system elements such as those processes specified in the standard, they may be excluded. This results in a reduction in the scope of certification. Reduced scope can only apply to the area of 'Product and/or Service realisation'. (Section 7 of the standard) which incorporates the design elements.

ISO 9001:2008 and ISO 9004 are a pair of quality management system standards designed to encompass the generic categories of hardware, software, processed materials and services. ISO 9001:2008 identifies the quality management system requirements for ensuring conforming service and product for certification. ISO 9004 provides the necessary guidance for all aspects of a quality management system designed to improve a firm's overall performance.

ISO 9001:2008 process model approach

A major component of ISO 9001:2008 is the adoption of a structure based on a process model, compared to the structure of the old versions, which were based on 20 quality system requirements. However, the 20 elements in the current ISO 9001:2008 are clearly identifiable within the process-based model.

The design and implementation of a construction organisation's quality management system is influenced by varying needs, particular objectives, the products and/or services provided, and the processes and specific practices employed. It is not the purpose of the International Standard to imply uniformity of quality management systems.

Organisations are not obliged to change the structure of their quality management system and/or its documentation to align with the structure of ISO 9001:2008. It is appropriate that any documentation of an organisation's quality management system should be defined in a manner that is appropriate to its unique activities.

The CD2 of ISO 9001 describes the process model approach as follows:

> Any activity or operation which receives inputs and converts them to outputs can be considered as a process. Almost all product and/or service activities and operations are processes ...
>
> For organisations to function, they have to define and manage numerous interlinked processes. Often the output from one process will directly form the input into the next process. The systematic identification and management of the various processes may be referred to as the 'process approach' to management ...
>
> This International Standard encourages the adoption of the process approach for the management of the organisation and its processes, and as a means of readily identifying and managing opportunities for improvement.
>
> (ISO 1999b)

Figure 3.1 provides a pictorial representation of the process model.

The following section of text identifies the main clauses and requirements of ISO 9001:2008.

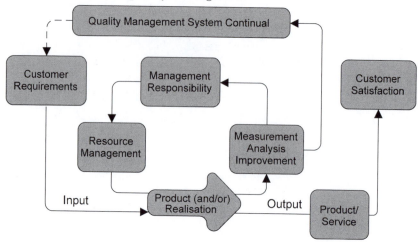

The main clauses and requirements of ISO 9001:2008

Figure 3.1 The revised quality management process model of ISO 9001:2000 (ISO Vol. 8, No. 3 May/June 1999a)

The starting point is Section 4 – Quality Management System. The previous pages relate to contents, foreword, introduction and other information as follows:

1. Scope

1.1 General
1.2 Application

2. Normative references

3. Terms and definitions

4. Quality Management Systems

4.1 General Requirements:

- establish and document your QMS
- implement your QMS
- maintain and improve your QMS.

The emphasis is placed on identifying and describing the processes involved rather than having formally documented procedures.

Has the construction organisation established, documented, implemented, maintained and continually improved the quality management system in accordance with the International Standard?

a) Has the construction organisation identified the processes needed for the quality management system and its application throughout the organisation?
b) Has the construction organisation determined the sequence and interaction of these processes?
c) Have they identified the criteria and method for effective operation and control of these processes?
d) Are the resources and information necessary to support the operation and monitoring of the process available?
e) Does the construction organisation measure, monitor and analyse the processes to achieve the planned results?
f) Does the construction organisation implement actions necessary to achieve planned results and the continual improvement of the process?
g) Does the construction organisation manage these processes in accordance with the requirements of this standard?

Where the construction organisation chooses to outsource any process that affects product conformity with requirements, the organisation must ensure it has control over such processes. This control will have to be identified within the quality management system.

4.2 Documentation Requirements:

- Develop QMS documents including your Quality Policy and measurable Quality Objectives.
- Prepare your Quality Manual.
- Have Procedures and Work Instructions to reflect what you do and how you do it.
- Control your QMS records.

A distinction is made between the documentation required by the standard and that documentation which is required by the construction organisation to control its processes.

Does the quality management system documentation include:

a) A quality policy and quality objectives?
b) A quality manual?
c) The documented procedures required in the International Standard?
d) Documents required by the construction organisation to ensure the effective planning, operation and control of its processes?
e) The quality records required by the International Standard?

Is there a quality manual containing the following:

a) the scope of the QMS including details and justification for exclusions?
b) documented procedures or references to them?

   c) a description of the interaction of the processes of the quality management systems?

Are relevant documents controlled?

Is there a documented procedure that ensures:

   a) Documents are approved for accuracy?
   b) They are reviewed and updated as necessary including re-approval?
   c) The current revision status and changes to documents are identified?
   d) Relevant versions of the documents are readily available at points of use?
   e) Documents are legible, readily identifiable and retrievable?
   f) Documents of external origin are identified and their distribution is controlled?
   g) There is a system to prevent the unintended use of obsolete documents and they are identified where they are retained for any purpose?

Are records required for the QMS established and maintained and is there a documented procedure available for this?

5. Management responsibility

5.1 Management commitment

Organisations must:

- Demonstrate their commitment to quality.
- Promote the importance of quality within their organisation.
- Support the development of their QMS.
- Support the implementation of their QMS.
- Support the continual improvement of their QMS.

Is there evidence of senior management commitment to development, implementation and improvement of the quality management system?

   Also has the construction organisation:

a) communicated to everyone the need to meet customer, statutory and legal requirements;
b) established the quality policy;
c) established quality objectives;
d) conducted management reviews;
e) ensured the availability of necessary resources?

   The quality policy only needs to be communicated and understood at appropriate rather than all levels of a construction organisation. The policy statement continues to be reviewed and becomes a controlled document.

Do senior management ensure that the quality policy:

a) is appropriate to the purpose of the construction organisation;
b) includes commitment to complying with requirements and to continually improve the effectiveness of the QMS;
c) provides a framework for establishing and reviewing quality objectives;
d) is communicated and understood within the construction organisation;
e) is reviewed for continual suitability?

5.2 Customer focus

This requires that construction organisations seek to:

- Enhance customer satisfaction by ensuring that you identify customer requirements.
- Enhance customer satisfaction by ensuring that customer requirements are being met.

5.3 Quality policy

Organisations must:

- Ensure that your organisation's Quality Policy serves its overall purpose.
- Ensure that your Quality Policy makes a commitment to continually improve the effectiveness of your QMS.
- Ensure that your Quality Policy supports your Quality Objectives.
- Ensure that your Quality Policy is communicated and discussed throughout your organisation.
- Ensure that your Quality Policy is periodically reviewed to make sure that it is still relevant to your organisation.

5.4 Planning

This requires companies to:

5.4.1   Establish and periodically review measurable Quality Objectives.
5.4.2   Plan the establishment, documentation and implementation of your QMS. Plan to ensure that you maintain and continually improve your QMS.

The responsibility for quality objectives must be established across the whole spectrum of the construction organisation's activities, including, of course, those needed to meet the requirements of the product. Quality objectives must be measurable.

Are quality objectives established at relevant functions and levels within the construction organisation?

Senior management are responsible for making plans and ensuring the objectives of the organisation are met.

Has senior management ensured that:

a) the QMS is planned in order to meet the quality objectives of the organisation;
b) the QMS is planned to meet the requirements outlined in clause 5.4.1;
c) the integrity of the QMS will be maintained when changes to the QMA are planned and performed?

5.5 Responsibility, authority and communication

Organisations must ensure that they:

5.5.1   Define and communicate your organisation's responsibilities and authorities.
5.5.2   Appoint a Management Representative from the organisation's management team.
5.5.3   Ensure that appropriate internal communications are established throughout your organisation.

Has senior management appointed a member of management with the required authority to:

a) ensure the processes of the QMS are established and maintained;
b) report to senior management on the performance of the QMS including the need for improvement;
c) promote awareness of customer requirements throughout the organisation?

Has senior management ensured that:

a) communication processes have been established within the construction organisation;
b) communication takes place regarding the effectiveness of the QMS?

5.6 Management review

Companies need to:

5.6.1   Review the organisation's QMS at planned intervals.
5.6.2   Examine management review inputs including records and data.
5.6.3   Generate management review outputs including decisions, actions and improvements.

Does senior management review the continuing suitability, adequacy and effectiveness of the quality management system?

a) Are the frequencies of reviews planned?
b) Does it assess opportunities for improvement?
c) Does it evaluate the need for changes to the QMS?
d) Does it include the quality policy and objectives?

Are records of management reviews maintained?

Do inputs to the management review include current performance and improvement opportunities and:

a) results of audits;
b) customer feedback;
c) process performance and product conformity;
d) status of preventative and corrective actions;
e) follow up actions from previous management reviews;
f) planned changes that could affect the quality management system;
g) recommendations for improvement?

Do outputs from management review include:

a) improvement of the quality management system and its processes;
b) improvement of product related to customer requirements;
c) resource needs?

6. Resource management

Companies should ensure:

6.1 Provision of Resources. Identify and provide resources required to implement, maintain and continually improve your QMS. Identify and provide resources required to enhance customer satisfaction by meeting customer requirements.

6.2 Human Resources

6.2.1 Employ and use competent personnel.
6.2.2 Regularly review employee competence. Provide and evaluate training and keep records of education, training, skills and experience. Promote employee awareness of quality within the construction organisation.

6.3 Infrastructure. Identify, provide and maintain infrastructure needs such as buildings, work areas, equipment and transport. Infrastructure includes hardware, software, and information systems and communications equipment.

6.4 Work Environment. Identify and manage the work environment required to carry out your activities satisfactorily. Examples of conditions under which work is performed could include safety, hygiene, noise, vibration, temperature, humidity, light, air quality, weather, etc.

Has the construction organisation determined the resources it needs and are they provided in order to:

a) implement and maintain the quality management system;
b) continually improve the effectiveness of the quality management system;
c) address customer satisfaction?

Personnel shall be 'competent', which carries a very specific meaning (this point relates back to the manpower planning issue previously noted).

Are personnel performing work affecting quality competent on the basis of education, training, skills and experience?

a) Have competency needs for personnel performing activities affecting quality been established?
b) Has training, or another alternative, been provided, where necessary, to satisfy these needs?
c) Has the effectiveness of the actions taken been evaluated?
d) Has management ensured employees are aware of the relevance and importance of their activities and how they contribute to achieving the quality objectives?
e) Has management maintained records of education, training, skills and experience?

Has management identified, provided and maintained the facilities it needs to achieve the conformity of product, including where necessary:

a) buildings, workspace and associated utilities;
b) process equipment, hardware and software;
c) supporting services (e.g. transport, communication)?

Does the construction organisation identify and manage the factors of the work environment needed to achieve conformity of product?

7. Product realisation

It is expected that companies shall ensure:

7.1 Planning of Product Realisation. Plan, develop and record product realisation processes.
7.2 Customer Related Processes

A requirement that organisations will:

7.2.1 Identify customers' product requirements including delivery and post-delivery activities. Verify product statutory and regulatory requirements and any other additional requirements. This could include warranties, maintenance, recycling, final disposal, etc.
7.2.2 Review your customers' product requirements, maintain records of product requirements, maintain records of product requirement reviews and control changes in customers' product requirements.
7.2.3 Communicate with your customers.

7.3 Design and Development

Companies need to:

7.3.1 Plan and design the development of your products. Control the design and development of your products. Update your planning outputs whenever product design and development progress makes this necessary.

This is the only section of the Standard where exclusions are permissible.

a) Are there quality objectives and requirements for the product?
b) Has management established processes, documents and resources specific to the product?
c) How does management verify, validate, monitor, inspect and test activities specific to the product and the criteria for acceptability?
d) Are there adequate records that provide evidence that realisation processes and resulting product meets requirements?

7.3.2 Identify product design and development outputs. Maintain a record of design and development inputs. Review your product design and development inputs.
7.3.3 Produce product design and development outputs. Approve product design and development outputs before they are formally released. Verify that product design and development outputs meet design and development input requirements.
7.3.4 Perform systematic design and development reviews throughout the design and development process. Maintain a record of design and development reviews.
7.3.5 Carry out design and development verifications. Maintain a record of design and development verifications.
7.3.6 Perform design and development validations. Maintain a record of design and development verifications.
7.3.7 Control design and development changes:

a) identify changes in design
b) record changes in design and development
c) review changes in design and development
d) verify changes in design and development
e) validate changes in design and development
f) approve changes in design and development before you implement these changes.

The aim is to establish the needs and requirements of the customer and this may be prior to receiving an order or contract. An emphasis is given to the implied needs of the customer and end users.
Does the construction organisation determine:

a)   The requirements specified by the customer, including the requirements for delivery and post-delivery activities, the needs for the availability and support of the product;
b)   Are requirements that are not specified by the customer but necessary for intended use identified;
c)   Obligations related to product, including statutory and regulatory requirements;
d)   What other requirements may it have in respect of the product?

Does the organisation review the requirements relating to the product?
Has the review to cover this been carried out before the decision or commitment to supply the product to the customer?

a)   Have product requirements been clearly defined?
b)   Have, where the customer provides no documented requirement, the offered requirements been confirmed with the client?
c)   Have contract or other requirements differing from those previously expressed been resolved?
d)   How have management confirmed that the construction organisation has the ability to meet the noted requirements?
e)   Have the results of the review and subsequent follow-up actions been recorded?
f)   When a product requirement has been changed has the construction organisation ensured all relevant documentation is amended?
g)   Have all the relevant personnel been advised of changed requirements?

Has the company identified and implemented effective arrangements for communicating with the customer?
Does it cover:

a)   Product information;
b)   Enquiry, contract or order processing, including amendments;
c)   Customer feedback, including customer complaints?

Further:

a) Have the stages of design and/or development processes been determined?
b) Have the review, verification and validation activities appropriate to each design and/or development stage been developed?
c) Has management clearly defined the responsibilities and authorities for design and/or development activities?
d) Have the interfaces between the different groups of design and/or development functions been managed to ensure communication is effective and responsibilities are clear?
e) Is planning output updated, as appropriate, during the progress of the design and/or development process?

Design and/or development inputs:

a) Have requirements for product function and performance been defined and documented?
b) Have the applicable statutory and legal requirements been defined and documented?
c) Is there applicable information derived from previous similar designs?
d) Have any other requirements essential for design and/or development been defined and documented?
e) Have these documented inputs been reviewed for adequacy? (Incomplete, ambiguous or conflicting requirements shall be resolved).

Design and/or development outputs

Are the outputs of the design and/or development process documented to ensure verification of the design and/or development inputs?

a) Design and/or development outputs shall meet the design and/or development input requirements.
b) Design and/or development output shall provide appropriate information for production and service provision.
c) Do design and/or development outputs contain or reference product acceptance criteria?
d) Does design and/or development define characteristics of the product that are essential to its safe and proper use?
e) Are design and/or development output documents approved prior to release?

Design and/or development review

Are systematic reviews of the design and/or development conducted at suitable intervals to:

a)   Evaluate the ability to fulfil note requirements;
b)   Identify problems and propose follow-up actions;
c)   Do the participants of the review include representatives of functions concerned with the design and/or development stage(s) being reviewed?
d)   Are the results of the reviews and the subsequent follow-up actions recorded?

Subsequent follow up action must be recorded and held as a quality record.

Validation is required in order to confirm that the resulting product meets the requirements for the intended use, as opposed to 'defined user needs or requirements'.

a)   Is validation undertaken to verify the resulting product is capable of meeting the requirements of the intended use?
b)   Is validation completed prior to delivery or implementation of the product?
c)   Where it is impractical to perform validation is it performed to the extent applicable?
d)   Are the results of validation and any follow-up actions recorded?

Control of design and or development changes

a)   Are design and/or development changes identified and recorded?
b)   Does it cover the evaluation of the effects of change on constituent parts and delivered products?
c)   Are the changes verified and validated, as appropriate, and approved before implementation?
d)   Are the results of the review and subsequent follow-up actions recorded?

7.4   Purchasing

Companies must:

7.4.1   Establish criteria that you can use to control suppliers. Evaluate your suppliers' ability to supply products that meet your organisation's requirements and select suppliers that are capable of supplying products that meet your organisation's specified requirements. Make sure that purchased products meet specified purchase requirements.

7.4.2   Describe your purchasing requirements. Ensure that purchasing requirements are adequately specified before you discuss them with suppliers.

7.4.3   Establish product verification or inspection methods in order to ensure that purchased products meet purchase requirements. Implement product verification or inspection methods in order to ensure that purchased products meet purchase requirements.

7.5   Production and Service Provision

Organisations are required to:

7.5.1   Carry out production and/or service provision under controlled conditions.

7.5.2   Validate production and service provision processes whenever process outputs cannot be measured, monitored, or verified until after the product is in use or the service has been delivered. Establish arrangements to control special processes.

7.5.3   Establish the unique identity of your organisation's products (if appropriate). Identify the monitoring and measurement status of your organisation's products.

7.5.4   Identify property supplied to you by customers and verify property supplied by customers. Protect and safeguard property supplied by customers. This includes work carried out on site on behalf of customers.

7.5.5   Make sure that your products and components continue to conform to requirements while they are being processed internally. Make sure that your products and components continue to conform to requirements while they are being delivered to the intended destination.

7.6   Control of Measuring and Monitoring Equipment

Companies have to:

- Identify your organisation's monitoring and measuring needs and requirements.
- Select equipment that can meet your organisation's monitoring and measuring needs and requirements.
- Establish monitoring and measuring processes.
- Calibrate your monitoring and measuring equipment whenever necessary to ensure that results are valid.
- Protect your monitoring and measuring equipment.
- Confirm that monitoring and measuring software is capable of doing the job you want it to do.
- Evaluate the validity of previous measurements whenever you discover that your measuring or monitoring equipment is out of calibration.
- Maintain calibration records.

The type and extent of control is dependent upon the effect it will have on subsequent processes and applies to all purchased items.

- Does the company control its purchasing to ensure purchased products conform to requirements?
- Is the extent of control appropriate to the effect on the subsequent realisation processes and outputs?

- Has the construction organisation evaluated and selected its suppliers based on their ability to supply products in accordance with requirements?
- Are the criteria for the selection of and periodic evaluation of suppliers defined?
- Are results of evaluation and follow-up actions recorded?

Purchasing information

Does the purchasing document contain information describing the product to be purchased, and including where appropriate:

a)   Requirements for the approval or qualification of product, procedures, processes, facilities and equipment;
b)   Requirements for qualification of personnel;
c)   Quality management system requirements?

Does the construction organisation ensure the adequacy of specified requirements contained in the purchasing documents prior to their release?

Verification of purchased product

Has the construction organisation established and implemented the inspection or other activities necessary for ensuring that purchased product meets specified requirements?

Has the construction organisation specified the intended verification arrangements for the product where the verification takes place at the supplier's premises?

Are the intended verification arrangements and method of product release included in the purchasing information?

Product and service provision control

Does the construction organisation carry out production and service provision under planned and controlled conditions, including, as applicable:

a)   The availability of information that specifies the characteristics of the product;
b)   The availability of work instructions (where necessary);
c)   The use of suitable equipment for production and service provision;
d)   The availability and use of measuring and monitoring devices;
e)   The implementation of monitoring activities;
f)   The implementation of defined processes for release, delivery and applicable post-delivery activities?

Note – The approval of processes (within production) and equipment and criteria for workmanship are not specified.

Validation of special processes

Has the construction organisation validated any production and service processes where the resulting output cannot be verified by subsequent measurement or monitoring?

Is there provision for processes where deficiencies may become apparent only after the product is in use or the service has been delivered?

Does validation demonstrate the ability of the process to achieve planned results?

Does the construction organisation define arrangements for validation that includes, where applicable:

a)   Approval/review of processes;
b)   Approval of equipment and qualification of personnel;
c)   Use of defined methods and procedures;
d)   Requirements for records;
e)   Re-validation?

Identification and traceability

The identification should not only define the product but also indicate its status in terms of inspected, tested, acceptable or unacceptable.

Has the construction organisation identified, where appropriate, the product throughout the product realisation?

Has the status of the product been, with respect to measurement and monitoring requirements, identified throughout the process?

Has the construction organisation controlled and recorded the unique identification of the product, where traceability is required?

Customer property

Customer property is an item or material that the customer wants the organisation to incorporate into the product or service being provided by the organisation. Management may decide that a procedure is required for the control of customer property, particularly if intellectual property is involved and specific security measures are required.

• Does the organisation deal with customer-supplied products?
• Does the organisation exercise proper care over customer property?
• Is it identified, verified, protected and maintained properly prior to incorporation into the product?
• Is there a process available for dealing with damage or otherwise unsuitable customer-supplied products?
• Does the customer supply any intellectual property and if so is it properly controlled?

Control of measuring and monitoring devices

This is a requirement to identify the measurements to be made in order to establish product conformity.

Has the construction organisation established processes to ensure that measuring and monitoring activities are capable of and are carried out in a manner that is consistent with the monitoring and measurement requirements?

Have they identified suitable measuring devices to assure conformity of product to specified requirements?

When required to maintain valid results, are measuring and monitoring devices:

a)   Calibrated or verified at specified intervals or prior to use, against measurement devices traceable to National/International Standards and adjusted where and as necessary;
b)   Safeguarded from adjustments that would invalidate the measurement result;
c)   Protected from damage and deterioration during handling, maintenance and storage;
d)   Results of calibration recorded;
e)   Checked for validity of previous results and where necessary corrective action taken?

8 Measurement, Analysis and Improvement

This section requires that firms:

8.1   General. Identify the monitoring, measurement, and analytical processes that your organisation needs to have in order to be able to demonstrate conformity and make improvements. Plan how monitoring, measurement, and analytical processes will be used to demonstrate conformity and make improvements. Implement your organisation's monitoring, measurement, and analytical processes.
8.2   Monitoring and Measurement

There is a need to:

8.2.1   Establish methods that you can use to monitor and measure customer satisfaction (perceptions).
8.2.2   Establish an internal audit procedure and carry out internal audits of your QMS and take appropriate actions to address results.
8.2.3   Select suitable methods to monitor and measure the processes that make up your organisation's QMS. Take appropriate action whenever your QMS processes fail to achieve planned results.
8.2.4   Monitor your organisation's product characteristics. Measure your organisation's product.

8.3   Control of Nonconforming Product. Establish and document a procedure to identify and control nonconforming products. Identify and control your nonconforming products. Re-verify nonconforming products that were corrected. Control nonconforming products after delivery or use. Maintain records of nonconforming products.

8.4   Analysis of Data. Define quality management information needs and collect QMS performance data. Provide information for management by analysing data and utilise the findings to develop measurable Quality Objectives.

8.5   Improvement

It is necessary to:

8.5.1   Continually improve the effectiveness of your QMS and use information and data to improve the effectiveness of your QMS.

8.5.2   Establish a corrective action procedure and document your corrective action procedure.

8.5.3   Establish a preventative action procedure and document your preventative action procedure. Implement and maintain your corrective action procedure.

Has the construction organisation planned and implemented monitoring, measurement, analysis and continual improvement monitoring activities? These need to:

a)   Demonstrate conformity of the product;
b)   Assure conformity of the quality management system;
c)   Achieve improvement of the effectiveness of the quality management system?

Does it determine the need for and use of applicable methodologies (including statistical techniques)?

Customer satisfaction

Does the construction organisation monitor information on customer perceptions of whether or not the organisation has met customer requirements?

Are the methodologies for obtaining and using information determined?

Has the construction organisation determined the customer satisfaction data it needs to collect?

Internal audit

Personnel other than those who actually perform the activity being audited can conduct audits. This is particularly useful to small organisations where finding an independent auditor is difficult. Internal audits must also confirm conformance to the Standard. Internal audits previously focussed on compliance with internal procedures.

Do the audits check that the QMS complies with the planned arrangements, quality management system requirements and the requirements of the Standard?

Do the audits check that the QMS has been effectively implemented and maintained?

Are the audits planned taking into account the status and importance of the area being audited?

Do audits have a clear scope and do they take into account the previous audit findings when setting the frequency and methodology?

Are audits conducted by personnel whose choice will ensure the objectivity and impartiality of the audit process?

Does the process ensure that auditors do not audit their own work?

Is there a procedure covering audit requirements and does it define the responsibilities and requirements for conducting audits ensuring their independence?

Are results recorded and reported to management?

Does management take timely corrective action on deficiencies found during the audit and do they follow up actions verifying the implementation of corrective actions? Is this reported and recorded?

Measurement and monitoring processes

This requirement draws on statistical techniques.

- Are there suitable methods for measuring and monitoring the processes necessary to meet customer requirements?
- Do these methods confirm the continuing ability of each process to satisfy its intended purpose?

Measurement and monitoring of product

- Are the characteristics of the product measured and monitored to verify that the requirements for the product are met?
- Are these checks carried out at appropriate stages of the process?
- Is there documented evidence of conformity with the acceptance criteria?
- Do records indicate the authority for release of the product?
- Is it clear that product release cannot proceed until all specified activities have been completed satisfactorily?
- Is there a process by which concession approval by the customer may be obtained?

Control of nonconformity

A documented procedure is required for controlling nonconformity in accordance with the requirements. Also:

- Are products that do not conform identified as such?
- Are they prevented from unintended use or delivery?
- Are these processes defined and documented in procedure?
- Are nonconforming products corrected and subject to re-verifications to demonstrate conformity?
- When nonconformity of product is detected after delivery or use does the organisation take appropriate action regarding the consequences?
- Is it clear where rectification of a nonconforming product needs to be reported to the customer, end-user, regulatory body or other?

Analysis of data

Does the construction organisation determine, collect and analyse appropriate data to determine suitability and effectiveness of QMS?
Does this identify where continual improvements can be made?
Does this data include measuring and monitoring activities etc?
Does the construction organisation analyse the data to provide information on:

a) Customer satisfaction;
b) Conformance to product requirements;
c) Characteristics of processes, products and their trends;
d) Suppliers?

Continual improvement

This is an explicit requirement which permeates the ISO 9001:2008 Standard.
   Does the construction organisation facilitate the continual improvement of the QMS by use of its:

a)   Quality policy;
b)   Objectives;
c)   Audit results;
d)   Analysis of data;
e)   Corrective and preventative actions;
f)   Management review?

Corrective action

Is there evidence that the construction organisation has taken corrective action to eliminate the cause of nonconformities?
Does this action prevent recurrence?
Is it appropriate to the impact of the problems encountered?

Does the documented procedure for corrective action define requirements for identifying nonconformities (including customer complaints):

a) Determining the causes of nonconformity;
b) Evaluating the need for actions to ensure that nonconformities do not recur;
c) Determining and implementing the corrective action needed;
d) Recording results of action taken;
e) Reviewing corrective action taken?

Preventative action

Has the construction organisation determined preventative action to eliminate causes of potential nonconformities to prevent occurrence?
Is preventative action appropriate to the impact of the problem?
Is there a documented procedure?
Does the documented procedure for preventative action define requirements for:

a) Determining potential nonconformities and their causes;
b) Evaluating the need for action to prevent occurrence of nonconformities;
c) Determining and implementing required action;
d) Recording results of action taken;
e) Review of preventative action taken?

(Watson and Watts 2000)

Organisational structure

The structure a construction company adopts must allow for the utilisation of the previously stated quality procedures and practices. However, managers should take note of the following, as advocated by Kanter:

Restructuring the company can entail threats to current productivity such as:

- The cost of confusion – people can't find things, they don't know their own telephone extension …
- Misinformation – communication is haphazard … Rumours are created and take on a life of their own.
- Emotional leakage – managers are so focussed on the tasks to be done and decisions to be made that they neglect or ignore the emotional reactions engendered by change. But the reactions leak out in other ways, sometimes in unusual behaviour.
- Loss of Energy – any change consumes energy – especially if the restructuring is perceived negatively.

- Loss of Key Resources – some companies handle consolidation in bureaucratic rather than human ways by establishing uniform policies.

(Kanter 1989)

However, the above accepted, it may still be necessary for some restructuring of the construction organisation to take place.

The lack of business success [and quality conformance] in many cases can be attributed to a persistence with an out-dated organisational structure in almost entirely new circumstances to which the organisational form is no longer capable of responding both efficiently and effectively.

(Pilcher 1986)

The inference from Pilcher's writings is that performance is linked to structure. The performance stated here is not only the total organisational performance but also that of the individual manager. This fact is corroborated by Mitzberg (1983): 'Just keeping the structure together in the face of its conflicts also consumes a good deal of the energy of top management.'

There may well be some resistance to change within the organisation when deploying a new QMS. Bowman and Asch (1989) identify the following factors influencing perception of change and responses:

- Change Factors – content and effect of change, speed and method of implementation.
- Personal Factors – general attitudes, personality, self confidence, tolerance of ambiguity.
- Group Factors – group norms, group cohesiveness, superior's reaction.
- Organisational Factors – change history, organisational structure and climate.

Most of the resistance can be overcome by allaying the employee's fear of change; this can be done by arranging briefing sessions for all concerned and keeping all informed as to the organisational and personal benefits of a QMS deployment.

When considering the type of organisational structure available, an overriding comment from Murdick et al. (1980) should be considered. 'Is the form the most effective from the viewpoint of both strategy and prompt response to competition'. Further one must not forget that 'The organisation must be capable of operating in a dynamic operational environment [especially in construction]. Structures are no longer viewed as a rigid definition of hierarchical levels' (Kilmann 1985).

Once the strategic choice has been made the next problem to be addressed is how to arrange the construction organisation's resources in order to achieve its stated mission, the successful implementation of a QMS, in line with other key strategic objectives.

Remember that the

> Types of resources include financial, technical material, informational and human because of the limitations of human beings, all the resources of a firm must be subdivided into manageable portions. To move these resources into action, the strategic choices must also be broken down into sub-sets of more specific objectives and tasks.
>
> (Kilmann 1985)

> The four essential building blocks for an organisation's structure are job descriptions, the structure of working relationships, decision-making complexes and operating procedures
>
> (Torrington and Weightman 1985)

All covered by ISO 9001:2008.

When the construction organisation has decided upon the requirements of the above-noted four points of consideration the development of the structure should follow fairly easily.

There are various types of organisational structure available for selection by construction organisations. Though many organisations may use a hybrid of the classical forms, senior managers must not forget that: 'An organisation cannot function without communication. Communications tie together the component parts of an organisation.' (Chilver 1984)

The structure must be conducive to strategy. 'Organisations are full of surprises because they are so hard to predict' (Bolman and Deal 1984). This is true because of the human element.

> An organisation is a social system deliberately established to carry out some definite purpose. It consists of a number of people in a pattern of relationships ... Every organisation has a program – a set of planned activities that can go well or badly ... The manager of an organisation is the person who has the primary responsibility for making its activities go well.
>
> (Glassman 1978)

If construction managers are going to meet the challenge stated by Glassman (above) they must be skilled in assigning individuals to tasks and functions and at integrating these individual activities into a group effort. In return this effort is dependent upon the organisational structure. Although construction organisations differ considerably in size, purpose and complexity, each develops some form of structure.

In undertaking responsibility for the purposeful design of an organisation the manager(s) must decide how much specialisation and co-ordination is necessary, in order to attain the advocated benefits of ISO 9001:2008 implementation.

Process approach

ISO 9001:2008 requires construction organisations to have a quality manual, which includes the documented procedures or references to them. It also has to include a description of the sequence and interaction of the processes that make up the quality management system. The scope of the system has to be defined, including the basis for the use of the 'Permissible Exclusions' clause.

Overall, the effect of the requirements of the Standard is to reduce the instances where documented procedures are mandatory and to allow construction organisations the freedom to determine the type and extent of documentation needed to support the operation of the processes that make up the quality management system.

However, procedures provide the means of monitoring and controlling the process and process control will need to be evident at various stages. Juran (1988) writes, 'Process control can take place at several stages of progression including set up (start up) control, running control, product control, facilities control.'

Process control in a service can be somewhat different from that in manufacturing and it is vital that all discrete stages are identified, so that no operation is omitted. This is due to the fact that in most service industries any error in service provision will have an 'instantaneous' impact on the customer; the term used is 'simultaneity'.

Management

To meet this requirement of ISO 9001:2008 the people who create policy, steer and direct the organisation will have to show how they participate in the development and direction of the QMS. This may be direct involvement in the process, attendance at meetings, delivery of presentations and communication sessions or any other activity that shows leadership with respect to the QMS.

Formal authority of a manager is when the authority is viewed as originating at the top of a construction organisation's hierarchy and flowing downward through the hierarchy via delegation. Informal authority is the right conferred upon the manager and their subordinates. Butler (1986) notes that 'I recognise that hierarchies are essential.' However, the real source of authority possessed by an individual lies in the acceptance of its exercise by those who are subject to it. Formal authority is therefore in effect nominal authority (Herbst 1976).

It is senior management's responsibility to explain to everyone in the construction company that the system's approach to quality is not a stick with which management shall beat them. It is a tool which can aid them in their operational environments. After all if people are given responsibilities to perform, 'Management control focuses on the activities of the responsibility centre' (Anthony 1988).

Quality policy and objectives

Generally the responsibility for policy formulation rests with the highest level in the construction organisation. However, 'There is considerable controversy as to whether policy is basically concerned with setting the goals of an organisation or with establishing a system of rules subject to which goals will be achieved.' The policy statement will incorporate objectives and

> Organisational objectives give direction to the activities of the group and serve as media by which multiple interests are channelled into joint effort. Some are ultimate and broad objectives of the firm as a whole; some serve as intermediate goals or subgoals for the entire organisation; some are specific and relate to short term aims.
>
> (Massie 1989)

The quality policy is the driving force of the system and commits the construction organisation both to meeting stated requirements and to improvement. This will become one of the key documents against which the performance of the quality system will be judged. The translation of the quality policy into practice is then facilitated by the definition of supporting objectives. Quality objectives are now a clear requirement in their own right as opposed to a part of quality policy. Management must identify its policy upon the quality issues. They must be established widely within the organisation, support the policy, be measurable and focus on both meeting product requirements and achieving continual improvement.

Senior management may set the policy and objectives for the company but senior management should not forget that the deployment issues involve everyone in the organisation, especially the quality issues. People within construction firms are the mechanism of implementation.

Quality planning

Quality planning functions at two levels. At the senior management level it is their responsibility to ensure that the planning of the QMS, achievement of continual improvement and the setting of quality objectives takes place. At a lower level the organisation's quality documentation in relation to planning for the realisation of its processes is mandatory. Although the format is optional, quality plans only need to be as complex as the product or service demands.

Legal requirements

ISO 9001:2008 makes it clear that in determining customer needs and expectations this activity must include applicable regulatory and legal requirements. Compliance with such requirements is then invoked throughout contract review, design, and process control, etc.

Training and competence

Management should ensure that all employees are trained so that they may perform and implement the stated company objectives. New employees must also be made aware of the company objectives and trained if necessary. The emphasis is clearly on competence rather than just training. The comprehensiveness of training is dependent upon the company and should be embraced within an effective human resource strategy. A construction company must establish its training requirements for personnel. After identifying the requirements it is necessary to plan and implement training programmes. All training achievements should be recorded so that records may be updated and gaps in training and competences established and addressed. After all it is correct that 'Each person must know the content of his or her job and be trained to accomplish it' (Crosby 1990).

Information and communication

The Standard specifically requires organisations to ensure effective internal communication between functions regarding system processes and external communication with customers. This applies not only at the contract stage but also with respect to the provision of product information and in the obtaining of feedback. And of course 'For information to be useful it has to be accurate, valid and timely' (Bedworth and Bailey 1982).

Customer perception

Customer perception is addressed and requires sufficient information on satisfaction or dissatisfaction to be gathered in order to enable the construction organisation to monitor customer perception on whether or not customer requirements are being satisfied. Having no complaints may only mean that the construction organisation has no information, not that customer satisfaction has been achieved.

Analysis of data for improvement

This has been separated out from the body of corrective and preventative action, to plan and operate the system to facilitate the achievement of improvement is a specific requirement and must be clearly demonstrated.

Benefits of ISO 9001:2008 deployment

A summary of the theoretical advocated advantages of deployment to be obtained by construction companies includes the following:

ISO 9001:2008

- Provides a marketing focus for construction enterprises.
- Provides a means of achieving a top-quality performance in all areas/activities of the construction organisation.
- Provides clear and valid operating procedures for all staff.
- Critical audits are performed allowing for the removal of non-productive activities and the elimination of waste and hence non-value adding activities.
- Provides a corporate quality advantage, acting as a corporate competitive weapon.
- Develops group/team spirit within the company and thus leads to enhanced staff motivation.
- Improvement of corporate communication systems within an organisation.
- Reduced inspection costs, and hence improved corporate profitability.
- More efficient and effective utilisation of scarce resources.
- Recognition of Certification, leading to the possibility of obtaining more work.
- Customer satisfaction, i.e. providing the required customer quality every time and hence attaining possible re-engagement.

The above provide a very strong case for construction companies to seek certification to ISO 9001:2008; however, it may well be that a prospective client has a requirement that whomever they are going to do business with has certification. In which case a company may not be invited onto a tender list without it. This in itself is a valid rationale for seeking certification.

At this stage the reader should be able to identify the constituent parts and the rationale for the application of ISO 9001:2008.

Question two for the reader

What are the benefits of ISO 9001:2008 deployment for both the host organisation and its stakeholders? An answer is provided at the back of the book.

Problematic issues associated with the implementation of ISO 9001:2008

This section of the chapter focuses on the practical problematic issues associated with implementing an ISO 9001:2008 quality management system.

The previous section has established the main advocated advantages of implementation. However, the implementation process can prove to be a most problematic one, and the following establishes the critical issues, and provides advocated solutions designed to assist in the implementation process for construction-related organisations.

It cannot be overstated that senior management support is the most vital element required for the successful implementation of ISO 9008:2008.

If senior management support is not forthcoming the quality facilitator/manager (the person charged with the implementation of the quality management system) could also face further problematic issues such as:

- A lack of adequate authority to get people fully engaged.
- Insufficient funding for the project, leading to inadequate resources being allocated.
- Lack of sufficient time being allocated for the project; remember this is a vital resource and has to be provided for.
- Resistance to
  (i) information gathering and documentation production stages;
  (ii) the implementation process during the project.

Total commitment (from senior managers) needs to be demonstrated through 'policies' and 'overt support'.

If construction organisations are to avoid problems appertaining to resource issues senior management must provide the necessary resources. Senior management must take an active role in both designing and implementing the quality system, with support coming from the very start of the project. Senior management can show this commitment through the development of organisational policies which involve them in designing and implementing the quality system.

The two most important resource issues are that of adequate funding for the project and the allowance of sufficient time for people to participate. Participation is necessary when the quality facilitator is gathering information for writing of the Quality and Procedures manuals. Participation of staff is also vital during the implementation phase of the project. It should be noted that time allocation and funding are not mutually exclusive. A lack of funds can mean that money is not available to release staff when participation is required. Issues of authority and overcoming resistance to change are also not mutually exclusive. 'If appropriate authority does not accompany Managerial responsibilities and duties, the manager's effectiveness within the organisation is impaired' (Glassman 1978).

Glassman claims that managers should be delegated sufficient authority to complete their allocated tasks. Senior management need to ensure middle managers are not asked to perform tasks for which they have not been given the necessary authority to complete. There may well be some resistance to change within the host organisation. Coalitions of resistance could develop and if they are linked to a power base they could impede the implementation process.

The quality facilitator should try and overcome resistance by allaying employees' fear of change.

Managers within the construction organisation must 'manage'. They should not abdicate their responsibility to the quality facilitator (or team) without providing adequate authority.

Liebmann (1983) offers support to the above. When he was part of a quality team implementing a quality system he found that senior managers were 'Charged as to design a process to empower employees but did not empower the team. The result was failure of the Quality Project.'

Even before the implementation process begins, staff need to be made aware of the benefits of certification. They need to be convinced that the introduction of a quality system is worthwhile and can provide advantages for them and for the organisation. It is, therefore, senior management's duty to echo the rationale for the advantages of certification. This is an important issue since 'people tend to have an in-built resistance to change'.

The co-operation of staff is vital for successful implementation and in order for them to co-operate two issues require consideration:

- staff have to want to co-operate;
- staff have to be allowed to co-operate.

If staff are not coerced into being co-operative they will make a greater contribution to the implementation process.

It can be concluded that senior management support is a vital component at all stages of the design and implementation process. If this support is not provided a successful outcome of the project will prove to be most problematic.

Incorrect corporate culture

Quality systems require a corporate culture based on trust and a desire to identify problems in order to eliminate them. The concept of empowerment is a key component of an effective quality corporate culture.

If a climate of distrust exists between senior management and the rest of the company the implementation of the standard is doomed to fail. Organisational culture dictates the way a business operates, and how employees respond and are treated. Organisational culture contains such elements as a guiding philosophy, core values, purpose and operational beliefs. It must be understood that just following documented procedures and complying with standards will not guarantee success. Only if the correct culture exists will the true benefits be attained. A culture based upon morphogenic principles is required.

Whoever is charged with the task of designing and implementing a quality system must have the total support of the organisation. This support involves not only senior management but also the employees (the people who perform the documented tasks). If those responsible for implementation can obtain this total support for the system then successful implementation is possible. An important part of obtaining total involvement is to inform people of what the system is all about and to keep them informed throughout the design and implementation process. Successful implementation of any quality model depends upon the co-operation of all the people who are involved with it.

The following is an overall generic strategy for quality assurance system implementation, and is of value to construction organisations embarking upon the deployment process:

- obtain support from the total organisation/stakeholders;
- set realistic objectives/quality checks, such as timescale for the implementation process;
- deployment is a project so plan and programme activities ahead of time (engage project management principles);
- maintain internal and external contacts with key personnel/stakeholders;
- establish a clear review/monitoring feedback process;
- be flexible and willing to sacrifice time and other resources to obtain improvements;
- do not expect a great improvement in the saving of resources immediately. Have realistic targets;
- use expert opinions and advice when necessary – you may have to engage external consultants;
- do not expect too great an immediate return on investment. Some improvement projects may have key benefits because they provide customer satisfaction and assist the long-term survival of the organisation.

A final point made by Beck and Hillmar (1986) is worthy of note:

> A manager needs to be clear with employees about their roles and responsibilities and the results expected in order for them to know what they are accountable for. While holding employees accountable for performance the manager must be accountable to them for support.

Figure 3.2 provides a generic flow diagram for construction companies wanting to deploy and hence gain the advantages of ISO 9001:2008 certification.

Associated costs of certification

Potential benefits deriving from establishing and maintaining a certified quality management system are not secured without costs to the construction organisation.

These costs are both direct and indirect and include:

Direct

- Developing the quality management system.
- Producing the quality documentation.
- Establishing the implementation system.
- Maintaining the internal audit system.
- Independent third-party assessment.

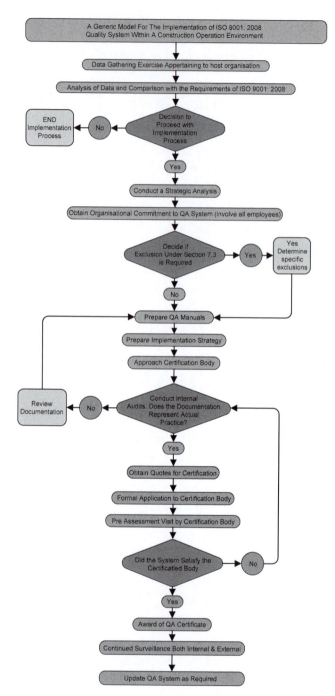

Figure 3.2 Model for the implementation of ISO 9001:2008

Indirect costs are difficult to assess but can include

- Liaising with the certification body.
- Changes to operational processes and procedures to accommodate certification requirements.
- Some demotivational aspects associated with staff and the implementation process.
- The consumption of organisational energy and efforts during the drive for certification.

There are also the costs of maintaining the system and surveillance visits by the certification body.

Certification bodies specify their various registration fees, which are subject to some variation depending on the following factors:

- The size of the company and number of employees.
- The structure of the organisation.
- Diversity and range of the company's activities.
- Nature and complexity of the quality system.
- Complexity of the documentation.

Conclusions

ISO 9001:2008 helps construction organisations by assisting them in obtaining a potential competitive weapon. Further it can enable the company to operate more efficiently and effectively, via investigating, in more detail, ways of improving performance in particular areas. For example ISO 9001:2008 focuses on processes and customer requirements.

Organisations moving to the new process model may need to rework their quality manual and move closer to TQM principles. At a superficial level, the model does more closely resemble the European Foundation for Quality Excellence Model approach (covered in Chapter 4). However, ISO 9001:2008 remains a process-based system with heavy emphasis on compliance. It also entails a rigorous external assessment of organisational performance, set against a standard, leading to accreditation. This may be required by potential clients before the company will be accepted on tendering lists.

Case study for the reader

The senior management of a construction company has been considering the deployment of ISO 9001:2008, as a means of being invited onto more potential client tender lists.

The managing director has always understood the value of quality as a potential competitive weapon, and is a quality advocate within the organisation.

The production of a quality provision is demanded by the company's clientele and this has been engrained in the organisation's workforce by the managing director.

But it has now become very clear to the company that both national and international clients require companies to hold ISO 9001:2008 certification before they are to be invited onto competitive tendering lists.

The managing director therefore decided to pursue the implementation of ISO 9001:2008 with vigour.

Based upon this chapter answer the following questions:

Question 1

Why may a potential customer demand that a supplier hold certification to ISO 9001:2008?

Question 2

The Managing Director decided to implement ISO 9001:2008. Why is it essential to have senior management support for the deployment process, and what are the likely outcomes if such support is not forthcoming?

Answers to the above questions are provided at the end of the book.

Further reading

Donnelly, M. (1999). Making the difference: quality strategy in the public sector. *Managing Service Quality*, 9 (1), pp. 47–52.

Hellard, R. B. (1993) *Total Quality in Construction Projects*. Thomas Telford, London ISBN 0 7277 1951 3.

Hutchin, T. (2001). *Unconstrained Organisations: Managing Sustainable Change*. Thomas Telford Ltd.

Naoum, S. (2001). *People and Organisational Management in Construction*. Thomas Telford.

Oakland, S. J. (1993) *Total Quality Management* Butterworth Heinemann, London.

References

Anthony, R. N. (1988). *The Management Control Function*. Boston: Harvard Business School Press.

Beck, A. C. and Hillmar, E. D. (1986). *Positive Management Practices*. London: Jossey Bass Ltd.

Bedworth, D. D. and Bailey, J. E. (1982). *Integrated Production Control Systems*. Chichester: J. Wiley.

Bolman, L. G. and Deal, T. E. (1984). *Modern Approaches to Understanding and Managing Organisations*. London: Jossey-Bass Inc.

Bowman, C. and Asch, D., (1989). *Strategic Management*. London: Macmillan Education Ltd.

British Standards Institution. (1999a). *Quality Management Systems – Fundamentals and Vocabulary*. London. ISO CD 2 9000, Draft.

——(2000a). *BSEN ISO 9000:2000 Quality Management – Fundamental and Vocabulary*, Milton Keynes.

——(2000b). *BSEN ISO 9001:2000 Quality Management Systems – Requirements*, Milton Keynes.

Butler, G. V. (1986). *Organisation and Management*, Prentice Hall UK Ltd.

Chilver, J. W. (1984). *People, Communication and Organisation*. Oxford, Pergamon Press Ltd.

Dale, B. G. (2004). *Managing Quality 4th ed*, Blackwell Publishing, Oxford, UK.

Dayton, J. B. (1988) *Quest for Quality: Developments in the Management of Quality by the United Kingdom*. Department of the Environment. Property Services Vol. 1. Agency. CIB W/65, Organisation & Management of Construction Proceedings. Department of the Environment.

Glassman, A. M. (1978). *The Challenge of Management*. Toronto: J. Wiley and sons.

Griffith, A. (1990) *Quality Assurance in Building*. London. Macmillan Education Ltd.

Herbust, P. H. G. (1976). *Alternatives to Hierarchies*. Netherlands, Mennen Asten.

Hogg, D. M. P. (1990). *The Enterprise Initiative, The Case For Quality*. London: Department of Trade and Industry.

International Organisation for Standardisation. *International Organisation for Standardisation*, 8. (1999a). No 3, May/June.

International Organisation for Standards. *International Organisation for Standards*. (1999b). http://www.iso,ch/9000@/summary.html.moy.

Johnson, G. and Scholes, K., (1984). *Exploring Corporate Strategy*. London: Prentice Hall International.

Juran, J. M. (1988). *Juran On Planning For Quality*. London: Macmillan Free Press.

Kanter, R. M. (1989). *When Giants Learn To Dance*. Reading: Cox Andwyman.

Kilmann, R. H., (1985) *Beyond The Quick Fix*. London: Jossey Bass Inc.

Liebmann, J. D. 1993. 'A Quality Initiative Postponed' in *New Directions For Institutional Research*. No 78, pp. 117–21.

Massie, J. L. (1989). *Essentials Of Management*. Englewood: Prentice Hall International.

Miles, R. and Snow, C. (1983). 'Some Tests Of Effectiveness and Functional Attributes Of Miles and Snow's Strategic Types' in *Academy of Management Journal*. 26(1) pp. 5–26.

Mitzberg, H. (1983). *Structure In Fives: Designing Effective Organisations'*. Englewood Cliffs: Prentice Hall.

Murdick, R. G. Eckhouse, Moor, R. C. and Zimmerer, T. W. (1980). *Business Policy, A Framework For Analysis*. Indianola: Grid Publishing Inc.

Peters, T. J. and Waterman, R. H. (1982). *In Serach of Excellence* Cambridge: Harper and Row.

Pilcher, R. (1986). *Principles of Construction Management*. Maidenhead: McGraw Hill.

The Institute of Quality Assurance (1990). *Training For Quality*. London: The Institute of Quality Assurance.

Torrington, D. and Weightman, J. (1985). *The Business of Management*. London, Prentice Hall UK Ltd.

Watson, P. and Watts, N. (2000). Quality Assurance – Complying with the New Standards. Construction Paper 114, *Construction Information Quarterly*. 2(1), 11–19.

4 The European Foundation for Quality Management Excellence Model (EFQM EM)

This chapter establishes the linkages between Total Quality Management (TQM) and the European Foundation for Quality Management Excellence Model (EFQM EM). The EFQM Excellence Model's constituent parts are established along with the advocated advantages of deployment and self-assessment methodologies. The chapter also provides a flow diagram designed by the book's authors, which should prove a useful tool for construction-related organisations engaged in the implementation process of the EFQM EM.

Learning outcomes

By the end of this chapter the reader will be able to demonstrate an understanding of:

- The linkages between TQM and EFQM EM.
- The constituent parts of the EFQM EM and its fundamental principles.
- The advantages of EFQM EM deployment.
- The application of the self-assessment methodologies and their part in the attainment of organisational improvement.

Introduction

This chapter explores the rationale for construction organisations engaging in the application of Total Quality Management (TQM). Further the European Foundation for Quality Management Excellence Model (EFQM EM) is explained and offered as a valid means of implementing TQM. The EFQM EM scoring system is demonstrated and its link to self-assesment and continuous improvement as part of a learning organisation culture are clearly established. After all, as noted by Oakland:

> The approach found in most world-class organisations is concerned with moving the focus of control from outside the individual to within, the objective being to make everyone accountable for their own

performance, and to get them committed to achieving [individual and organisational objectives] in a highly motivated fashion.

(Oakland 1997)

The EFQM Excellence Model was formerly known as the Business Excellence Model. It was developed between 1989 and 1991 by practitioners to bring together the various models for Total Quality Management (TQM) deployment.

European organisations were experiencing difficulties in the implementation of TQM principles, and hence attaining the following benefits of TQM application in practice:

- the production of a higher quality product/service through the systematic consideration of the client's requirements;
- a reduction in the overall process/time and costs via the minimisation of potential causes of errors and corrective actions;
- increased efficiency and effectiveness of all personnel with activities focused on customer satisfaction;
- improvement in information flow between all participants through team building and proactive management strategies.

TQM should assist in making effective use of all organisational resources, by developing a culture of continuous improvement. This empowers senior management to maximise their value-added activities and minimise efforts/ organisational energy expended on non value-adding activities.

TQM further enables companies to fully identify the extent of their operational activities and focus them on customer satisfaction. Part of this service focus is the provision of a significant reduction in costs, through the elimination of poor quality in the overall process. This empowers companies to attain a truly sustainable competitive advantage. TQM provides a holistic framework for the operational activities of enterprises. If a firm can overcome implementational problematic issues then a sustained competitive advantage is the reward to be gained (Watson and Chileshe 2001).

Some TQM proponents maintain that a common error in the application of TQM is the failure to recognise that every company, and environment, is different (Laza and Wheaton 1990, cited by Spencer 1994). Thus successful deployment is dependent upon the correct alignment of corporate strategies and operational environments, both encapsulated within a morphogenic culture.

Further, 'an important component of TQM is the implementation of work practices such as employee training, information sharing, involvement and empowerment' (Hendricks and Singhal 2001).

These are designed to produce a corporate culture where change and innovation are expected. They are to set 'corporate meanings' for employees and other stakeholders.

Brown (1998, cited by Ham 2005) defines organisational culture as 'A set of understandings or meanings shared by a group of people'. Thus Brown (1998) corroborates one of the advocated advantages for EFQM EM application.

The European Foundation for Quality Management Excellence Model (EFQM EM)

The application of TQM has been advocated by various eminent authors, for example Oakland (1993), Wright (1997) and Cherkasky (1992). However, the process of deployment can prove to be most problematic for many public- and private-sector organisations. Terms such as empowerment, cultural dynamics and cross-functional communications have only served to add to the confusion. Further, many Western organisations have endeavoured to adopt TQM based upon an Eastern philosophy and culture and this has complicated matters further. Therefore, a practical application for TQM deployment was required.

The European Foundation for Quality Management Excellence Model (EFQM EM) is a model for TQM deployment based upon practical application and feedback from practitioners; it is most suited for Western construction organisations.

EFQM Excellence Model development

The EFQM Excellence Model has been used extensively and beneficially in manufacturing, construction, banking and finance, education, management and consultancy. Companies apply the EFQM Excellence Model because the pursuit of business excellence through TQM is a decisive factor in allowing them to compete in today's turbulent global marketplace. EFQM is a non-profit-making organisation providing various networking, benchmarking and training events to help members keep up with the latest trends in business management and research in TQM. It launched the European Quality Award in 1991 to stimulate interest and it is awarded to those who have given 'exceptional attention' to TQM.

The function of the EFQM Excellence Model

Hillman (1994) suggested that 'the EFQM Model provided a tried and tested framework, an accepted basis for evaluation and a means to facilitate comparisons [for construction firms] both internally and externally'.

The British Quality Foundation is the UK sponsor of the EFQM Excellence Model. The aim of the British Quality Foundation is to promote continuous improvement and organisational excellence using the EFQM Excellence Model. It has over 1,700 member organisations. The philosophy of the Foundation is succinctly put in the following quote:

Regardless of sector, size, structure or maturity, to be successful, organisations need to establish an appropriate management system. The EFQM Excellence Model is a practical tool to help [construction] organisations do this by measuring where they are on the path to Excellence; helping them understand the gaps; and then stimulating solutions. The EFQM is committed to researching and updating the Model with the inputs of tested good practice from thousands of organisations [including construction organisations] both within and outside of Europe. In this way we ensure the model remains dynamic and in line with current management thinking.

(European Foundation for Quality Management 2000a)

The model was updated in April 1999 after widespread consultation with members. Significant enhancements were made and in particular they ensured the relevance of the model to the public sector. The enhanced model is officially referred to as the European Foundation for Quality Management (EFQM) Excellence Model.

Table 4.1 provides an overview of the basic philosophy and concepts of the EFQM Excellence Model.

The EFQM Excellence Model is a non-prescriptive framework for self-assessment. Using this tool a construction or any organisation can assess whether it is doing the right things and obtaining the required results. The ensuing assessment of an organisation's performance is measured both by results and by the quality of the processes and systems developed to achieve them. In its most sophisticated form the model is used to assess an organisation for quality awards – including the European Quality Award. The assessment encompasses the whole organisation (or the whole of a part of an organisation) using nine standard criteria. The model provides a balance and a relationship between approach (the way in which results are achieved) and results (what is achieved) – a balance between cause and effect. The criteria which deal with causes are known as enablers. Those which deal with effects are known as results. In scoring the organisation both have equal weighting.

The following key questions must be addressed by a construction company seeking to be the 'best', and appropriate answers provided.

Is the construction company striving for:

- Better products/services;
- Better organisation/management;
- Better information/communication systems?

The above are not mutually exclusive and a construction company may well be seeking all three.

Harrison (1993) sees the way forward as first creating a vision for the company, second installing ownership of the issues facing the company, and

Table 4.1 The philosophy behind the EFQM EM's basic concepts

Leadership & modelling	The behaviour of an organisation's leaders creates a clarity and unity of purpose and an environment in which the organisation and its people can excel.
Policy & strategy	A successful organisation formulates policy and strategy in collaboration with its people and it is based on relevant, up to date and comprehensive information and research.
Continuous learning & improvement	Organisational performance is maximized when it is based on the management & sharing of knowledge within a culture of continuous learning innovation & improvement.
Partnership development	Mutually beneficial relationships, built on trust, sharing of knowledge and integration with partner organisations are a crucial resource to any effective organisation.
Management by processes & facts	Organisations perform more effectively when all inter-related activities are systematically managed and decisions about current operations and improvements are based on reliable information and stakeholder perceptions.
Customer focus	Quality of service and retention of market share are best achieved through a clear focus on the current and potential needs of customers.
People development & involvement	The full potential of an organisation's people is best realised through shared values and a culture of trust and empowerment which involves everyone.
Public responsibility	The interests of the organisation and its people are best served by adopting an ethical approach and exceeding the expectations and regulations of the community at large.
Results orientation	Excellence depends on balancing and satisfying the needs of all relevant stakeholders including employees, customers, suppliers and society as well as the funding organisation.

third he advocates planning and implementation of a continuous change process. However, above all, construction companies must 'Focus on their customers'.

Some of the techniques and issues to be addressed by organisations in order to function at best practice level include the following:

The first issue to be considered is that of senior management commitment.

There is no substitute for effective leadership by senior management. The number of failures on the road to implementing TQM and thus operating at best practice level blamed on a lack of management commitment suggests that not only is it needed but that it cannot be assumed and may be difficult to obtain in practice (Nunney 1992). This important issue has previously been explored in more detail in Chapter 3 of this text. After all it is vital that 'People at all levels in the [construction] organisation must fully understand the organisational objectives and the timeliness of the objectives' (Mundy 1992). The above, as noted by Mundy, is a function of senior management activities. A further consideration for an organisation pursuing a strategy of best practice are the Four Ps: Purpose, Planning, Process and Performance measurements.

Purpose

People perform better when they understand the objectives of the company, and also appreciate that teamwork is an essential element, therefore people should be trained to work in teams. Organisations must embrace the concept of 'empowerment'. People need timely feedback upon their performance as this affects the intrinsic motivation of individuals.

Planning

Planning and monitoring requires the setting up of a dynamic closed feedback loop. A further requirement is that the senior management team (SMT) must not just think in terms of reducing labour costs per service facility or product manufactured. Consideration should be given to reducing the time taken for materials to undergo each production process. This of course requires planning, incorporated within which should be a collaborative partnership approach towards suppliers, with the attainment of long-term strategic benefits being the main objective.

Process

The process(es) must be flexible and this may demand as much organisational effort in support facilities as it does in operational activities. Design activities must be incorporated with the production function; this requires a holistic approach.

The process(es) will in most instances determine the service/product quality and such processes must add value. It should be appreciated by all concerned that quality is ultimately everyone's business.

Performance measurements

These require simple dynamic systems. However, it is not the quantity of data that is important but its usefulness in the management decision-making process. In summary it may be stated that the problematic issues are not the technological-based ones but the 'people issues'. 'Management acts to develop its people by caring for and training them' (Hickman and Silva 1989).

External and internal changes required of construction organisations

The concept of continuous improvement is an important aspect of a best practice performer and a fundamental function of TQM/EFQM EM.

(Yip 1992). Yip is advocating the dynamic nature of best-practice operations. In order for a construction organisation to become and/or remain a world-class performer it must consider both internal and external requirements.

Competition

> The trend towards economic liberalisation in general and privatisation in particular has had a major impact on business activity.
>
> (Preston 1993)

As noted by Preston the amount of competition for most construction organisations has increased, therefore companies must fully understand the nature of their competitive environment. Organisations must operate an all-embracing macro business strategy and not a micro one. Environmental scanning would provide the means for analysing a company's competitive environment.

Suppliers

The way a construction organisation deals with its suppliers must be as a joint venture(s), i.e. partnership(s). Construction companies and their suppliers must understand the synergistic advantages that are available to both parties and a win-win outcome for both parties must be their joint organisational goal.

Environmental factors

The environmental factors must be analysed to the extent that they provide either opportunities or threats to the construction company. The requirements of the environment may involve some physical alteration to the company. A Political, Economic, Social, Technological, Environmental and Legal (PESTEL) analysis would prove useful for organisations.

Economic factors

Construction companies are affected by and must respond to varying changes in monetary value. This factor will also impinge upon the supply and demand for the services or products subject to the rules of elasticity.

Technological factors

Considerable interest has been shown in the implementation of new technologies. This is because these technologies are not just for internal consumption but also enable construction organisations to communicate and interact on an international basis in real time.

The internal changes that may be required can be to a certain extent viewed as changing the production systems from the traditional 'push system' to a more modern 'pull system'.

Some of the issues raised under internal and external changes do overlap each other. However, one must not forget that the main internal change required is people orientated. 'Until recently, most senior operations managers did not perceive the human organisation as a source of competitive advantage' (Ross 1991).

Employees, as advocated by Kanter (1989), have an in-built resistance to change. Nevertheless, change may be necessary. For example, the type of organisational structure the company utilises can affect its ability to function at a best-practice level. Lack of business success in the deployment of TQM/EFQM EM in many cases can be attributed to a continued persistence with an outdated organisation structure.

If change is necessary then, as promoted by Ross (1991), 'The key to making the transition work is in the employee's understanding of its necessity [and value to them, as well as the company]'.

Usually groups of people in construction organisations recognise that work could be done more efficiently and/or effectively, but they are very rarely if ever asked for their opinion. It is worth noting the thoughts of Ross (1991): 'To participate effectively in the global marketplace, the implementation of production technologies must be combined with a programme aimed at aligning organisational structure and culture, the role and flow of information and people resources, if enterprises [including construction] are to exploit opportunities.'

The EFQM EM model provides effective linkages between people, processes and results, and thus addresses the valid points identified above.

EFQM EM functions

The EFQM state that the functions of their model may be split in four ways:

- as a framework which organisations can use to help them develop their vision and goals for the future, in a tangible and measurable way;
- as a framework which organisations can use to help them identify and understand the systemic nature of their business, the key linkages and cause-and-effect relationships;
- as the basis for the European Quality Award, a process which allows Europe to recognise its most successful organisations and promote them as role models of excellence from which others can learn;
- as a diagnostic tool for assessing the current health of the organisation.

Through this process a construction organisation is better able to balance its priorities, allocate resources and generate realistic business plans (European Foundation for Quality Management 2000a).

All these functions allow the model to be used for a number of activities, for example self-assessment, third-party assessment, benchmarking and as the basis for applying for the European Quality Award.

The constituent parts of the EFQM Excellence Model

The model incorporates business criteria, which are crucial to a construction company pursuing business excellence.

> More than ever, the EFQM Excellence Model offers an operational tool for the pursuit of excellence in performance and results.
> (European Foundation for Quality Management 1999b)

The model, which recognises that there are many approaches to achieving sustainable excellence in all aspects of performance, is based on the premise that:

Excellent results with respect to Performance, Customers, People and Society are achieved through Partnerships and Resources and Processes. The EFQM Model is presented in *Figure 4.1*.

The Excellence Model is based on the concept that both customer/people satisfaction and positive impact on society are achieved through leadership driving policy and strategy, people management, partnership and resources and processes leading ultimately to excellence in business results.

The EFQM Excellence Model provides firms with a way of achieving a top-quality performance. The 'enablers' and 'results', together with their

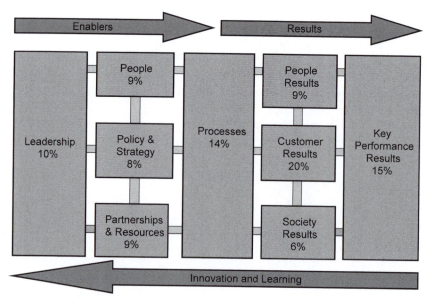

Figure 4.1 The EFQM Excellence Model

sub-criteria, provide a guide for construction organisations, suggesting which areas to focus on in order to succeed in satisfying their respective clients' requirements.

The EFQM Excellence Model is based on and supported by the specific concepts which are referred to as 'The Fundamental Concepts of Excellence'; these are:

- Results Orientation
- Customer Focus
- Leadership & Constancy of Purpose
- Management by Process & Facts
- People Development & Involvement
- Continuous Learning, Improvement & Innovation
- Partnership Development
- Corporate Social Responsibility

Each concept plays a role directly and indirectly related to different criterion and sub criterion within the EFQM Excellence Model. There are nine criteria which an organisation measures its performance against:

- Leadership (10%)
- People (9%)
- Policy and strategy (8%)
- Partnerships and resources (9%)
- Processes (14%)
- People results (9%)
- Customer results (20%)
- Society results (6%)
- Key performance results (15%)

The model as depicted in *Figure 4.1* consists of two main parts, **'Enablers'**, which are made up of five criteria, and **'Results'**, which consists of four criteria. Basically, the 'Enablers' deal with what an organisation does, while the 'Results' deal with what an organisation achieves. Most important is that mechanism of feedback within the model in which 'Results' are generated by 'Enablers' and 'Enablers' improved by using the feedback from 'Results' (EFQM 1999b).

Model contents and structure

The model's nine boxes in *Figure 4.1* represent the criteria against which to assess an organisation's progress towards excellence. Each of the nine criteria has a definition, which explains the high-level meaning of that specific criterion.

To develop the high-level meaning further each criterion is supported by a number of sub-criteria. Sub-criteria pose a number of questions that should be considered in the course of an organisation's assessment of its activities in relation to the model. Each criterion and sub-criterion is now explored in detail.

Leadership

DEFINITION

How leaders develop and facilitate the achievement of the mission and vision, develop values required for long-term success and implement these via appropriate actions and behaviours, and are personally involved in ensuring that the organisation's management system is developed, implemented and continually monitored.

SUB-CRITERIA

Leadership covers the following four sub-criteria:

- Leaders develop the mission, vision and values and are role models of a culture of excellence.
- Leaders are personally involved in ensuring the organisation's management system is developed, implemented and continuously improved.
- Leaders are involved with customers, partners and representatives of society.
- Leaders motivate, support and recognise the organisation's employees.

Policy and strategy

DEFINITION

How the organisation implements its mission and vision via a clear stakeholder-focused strategy, supported by relevant policies, plans, objectives, targets and processes.

SUB-CRITERIA

Policy and strategy covers the following five sub-criteria:

- Policy and strategy are based on the present and future needs and expectations of stakeholders both internal and external to the company.
- Policy and strategy are based on information from performance measurement, research, learning and creativity-related activities.

- Policy and strategy are developed, reviewed and updated.
- Policy and strategy are deployed through a framework of key processes.
- Policy and strategy are communicated and implemented. One must not forget that implementation is a process undertaken by people and therefore they have to be involved from the start of the process.

People

DEFINITION

How the organisation manages, develops and releases the knowledge and full potential of its people at an individual, team-based and organisation-wide level, and plans these activities in order to support its policy and strategy and the effective operation of its processes, all of which must be customer focused.

SUB-CRITERIA

People covers the following five sub-criteria:

- People resources are planned, managed and improved.
- People's knowledge and competencies are identified, developed and sustained in line with organisational activities.
- People are involved and empowered at all levels of the organisation.
- People within the organisation have an effective dialogue.
- People are rewarded, recognised and cared for.

Partnerships and resources

DEFINITION

How the organisation plans and manages its external partnerships and internal resources in order to support its policy and strategy and the effective and efficient management of partnerships and resources.

SUB-CRITERIA

Partnerships and resources covers the following five sub-criteria:

- Internal and external partnerships
- Finances
- Buildings, equipment and materials
- Technology
- Information and knowledge

Processes

DEFINITION

How the organisation manages its operational processes.

SUB-CRITERIA

Processes covers the following five sub-criteria:

- Processes are systematically designed and managed.
- Processes are improved as needed, using innovation in order to fully satisfy and generate increasing value for all stakeholders.
- Products and services are designed and developed based on customer needs and expectations.
- Products and services are produced, delivered and serviced.
- Customer relationships are managed and enhanced.

Customer results

DEFINITION

What the organisation is achieving in relation to its external customers.

SUB-CRITERIA

Customer results covers the following two sub-criteria:

- Perception measures
- Performance indicators

Both have to be incorporated into a closed feedback loop to ensure effective and efficient actions are taken ensuring continuous improvements.

People results

DEFINITION

What the organisation is achieving in relation to its people.

SUB-CRITERIA

People results covers the following two sub-criteria:

- Perception measures
- Performance indicators

This is a critical issue to be monitored in order to ensure full organisational participation in a drive for continual improvement.

Society results

DEFINITION

What the organisation is achieving in relation to local, national and international society as appropriate.

SUB-CRITERIA

Society results cover the following two sub-criteria:

- Perception measures
- Performance indicators

These results could be incorporated into the firm's social audit activities.

Key performance results

DEFINITION

What the organisation is achieving in relation to its planned performance.

SUB-CRITERIA

Key performance results cover the following two sub-criteria. Depending on the purpose and objectives of the organisation some of the measures contained in the guidance for key performance outcomes may be applicable to key performance indicators and vice versa. What the organisation is achieving in relation to its planned performance.

- Key performance outcomes
- Key performance indicators

These results must be examined in relation to the corporate vision and mission statements as incorporated in the corporate strategic plan.

The EFQM model provides a valuable framework for addressing the key operational activities of construction organisations. It is useful because it enables a link to be made between people, organisational objectives and improvement processes, all encompassed under the umbrella of continued improvement (EFQM 1999b).

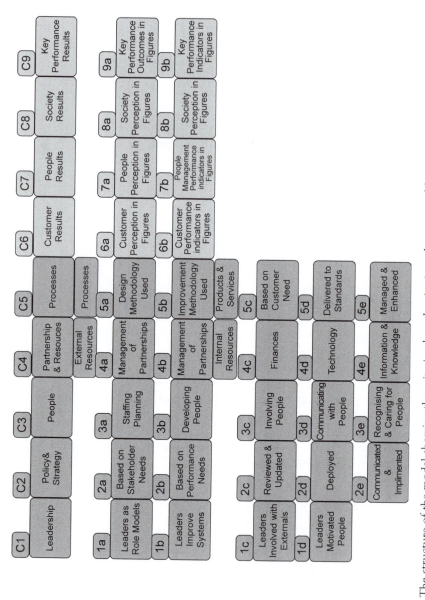

Figure 4.2 The structure of the model showing the criteria, the sub-criteria and areas to address

Cross-functional linkages of the EFQM Excellence Model

The following section identifies the cross-functional linkages of the EFQM EM. The integrative nature of the model is designed to provide first-level operational linkages between enablers and results.

First-level linkages are established in *Figure 4.2*.

Linkages between customer results and enablers

Customer results will be influenced and affected by the following enablers sub-criteria as illustrated in *Figure 4.2*.

- Leadership
- Sub-criterion 1C: Leaders' involvement with customers.
- Policy and Strategy
- Sub-criterion 2A: Establishing customers' needs and expectations.
- Sub-criterion 2C: Balancing customers' needs and expectations.
- People
- Sub-criterion 3B: People have the skills and competencies to deal with customers.
- Sub-criterion 3C: People's involvement with customers.
- Partnerships and Resources
- Sub-criterion 4A: Good supplier/partner relationships to satisfy customers.
- Processes
- Sub-criterion 5B: Improving process to satisfy the customer.
- Sub-criterion 5C: Product and service development.
- Sub-criterion 5D: Product and service delivery.
- Sub-criterion 5E: Customer relationship management.
- Figure 4.3 Linkages between customer results and enablers

Leadership

Excellent leaders develop and facilitate the achievement of the mission and vision and they develop organisational values and systems required for sustainable success and implement these via their actions and behaviours. During periods of change they retain a constancy of purpose. Where required, such leaders are able to change the direction of the organisation and inspire others to follow.

SUB-CRITERIA

1a. Leaders develop the mission, vision, values and ethics and are role models of a culture of excellence.

1b. Leaders are personally involved in ensuring the organisation's management system is developed, implemented and continuously improved.

1c. Leaders interact with customers, partners and representatives of society.
1d. Leaders reinforce a culture of excellence with the organisation's people.
1e. Leaders identify and champion organisational change.

Oakland et al. (2002) emphasised that a key role of leadership was being the driving force and motivational tool for all employees. Leaders are involved in all activities and processes of an organisation. As a matter of fact, the EFQM Excellence Model clearly shows such an importance by linking the leadership role with all the other criteria as depicted in *Figure 4.4*.

Wright (1997) emphasised that managers and employees in public organisations must consider that they 'cannot control the governmental and organisational decisions but they can change their reactions to them'. In addition, Wright (1997) pointed out that the role of a director must be managing the system rather than managing individuals within the system.

The importance of leadership

The leadership role is a critical function within the application of the EFQM Excellence Model; 'leadership drives policy and strategy through management of people, processes and resources, to deliver superior people satisfaction, customer satisfaction and societal satisfaction, which in turn leads to better business results', a requirement for all construction firms (EFQM 1999b).

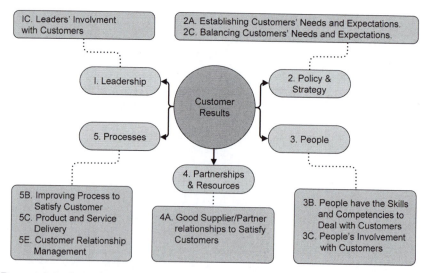

Figure 4.3 Linkages between customer results and enablers

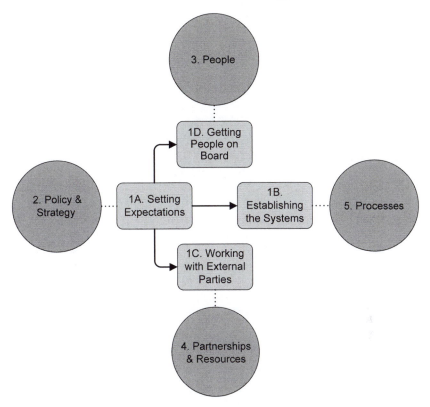

Figure 4.4 Linkages within the Leadership criterion

Policy and strategy

Excellent construction organisations implement their mission and vision by developing a stakeholder-focused strategy that takes into account both the market and sector in which it operates. Policies, plans, appropriate objectives and processes are developed and deployed to deliver the strategy. *Figure 4.5* depicts the extent of the linkages.

Sub-criteria

2a. Policy and Strategy are based on the present and future needs and expectations of stakeholders.
2b. Policy and Strategy are based on information from performance measurement, research, learning and external related activities.
2c. Policy and Strategy are developed, reviewed and updated.
2d. Policy and Strategy are communicated and deployed through a framework of key processes.

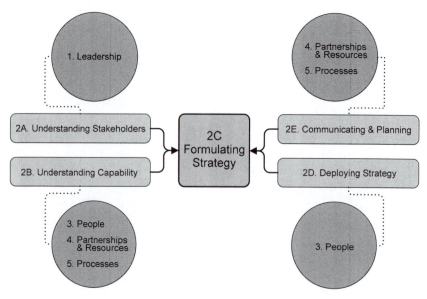

Figure 4.5 Linkages within the Policy and Strategy criterion

Definition of strategy

Orsini (2000) offers simple and succinct definitions of the following related to strategy:

- Vision: is what an organisation wants to be.
- Strategy: is the plan by which an organisation will attain the vision.
- Policies: outline the course of actions that will lead to the plan being carried out.

Thus, for the attainment of success in excellence and TQM in construction organisations, all three must align and reinforce one another. In other words, these definitions have to complement and interrelate with each other. Policies are often formed to deal with day-to-day operations. Johnson and Scholes (1999) advocate that the strategy of an organisation is shaped by several factors:

- environmental forces affecting the organisation;
- organisational resource availability;
- the values and expectations of senior management of the company;
- organisational stakeholder needs and wants.

Based on these factors, Johnson and Scholes (1999) defined strategy as ' ... the direction and scope of an organisation over the long term: which achieves advantage for the organisation through its configuration of resources within a changing environment, to meet the needs of markets and to fulfil stakeholder expectations'.

Types of strategies

There are three main types of strategy that can be found within organisations. The first type is 'corporate strategy', which can be characterised and described as ' ... the overall purpose and scope of the organisation to meet the expectations of major stakeholders and add value to the different parts of the enterprise'.

The second type is 'business unit strategy', which is mainly related to how a construction firm can compete successfully in its competitive marketplace. Since corporate strategy incorporates decisions related to the organisation as a whole, strategic decisions within this type of strategy would be related to a 'strategic business unit'. In a public-sector organisation the business unit will be ' ... a part of the organisation or service for which there is a distinct client group'. Thus, this type of strategy would cover the strategic decisions of how the needs of clients (stakeholders) may be met.

The third type of strategy is 'operational strategy', which can be summarised as how all resources, processes and employee skills can be integrated to carry out and accomplish the corporate and business strategic objectives in line with the corporate mission and vision (Johnson and Scholes 1999).

The gradual reframing of an organisation

A vital aspect by which the implementational effectiveness of an EFQM EM/ TQM framework model can be improved is through the gradual reframing of an organisation. These improvements would occur through determined phases; at the outset they should be well planned in order to address the expected consequences of cultural organisational change, which is never easy to achieve (Dooley 1998).

In addition, to be able to support such a culture change as a result of the introduction of the EFQM Excellence Model, there has to be demonstrable senior management involvement. It is evident from previous documented experiences that holding regular team meetings could be a good means for securing senior management support. These meetings provide a communication link between senior management, employees and middle management. After all, it has already been established that one of the most important barriers to TQM (and hence EFQM EM) success is insufficient management support (Capon et al. 1995).

What is evident is that the EFQM EM does encompass all construction organisational activities, including their impact upon society, and provides a valid methodology for construction organisations to obtain a sustainable competitive advantage.

Steps for implementing the EFQM Excellence Model

Ho (1999) summarised the critical steps for formulating and shaping a corporate strategy while undertaking the process of implementing the EFQM Excellence Model. The following steps relate to the relationship between a quality initiative and corporate strategy. Ho divided the corporate strategy into three key phases.

- The first phase is 'the determination of a corporate mission statement which sets a common value for everyone in the organisation', noting that a mission statement for an organisation is usually for a long-term period of at least 10 years.
- The second phase 'is defining the strategic options and choosing the optimum one'. This is the medium-term plan, which usually ranges from 3–5 years.
- The third phase is 'the strategic implementation which is also known as operations management'. This is the short-term plan, which usually lasts for three months to one year.

Having identified the above phases, Ho asked the question ' ... where does this TQM initiative – EFQM Excellence model – fit into this Corporate Strategy?' In order to address this issue it is best to consider quality as a routine organisational activity encompassed within the strategic planning and deployment process. The main advantage of this approach has been summarised by Ho, who notes that ' ... it adds totality to quality, as it is communicated throughout the [construction] organisation and spanned over its long term plan'.

Moreover, Ho has discussed the basis and rationale behind the success of TQM in Japanese companies. According to studies conducted on Japanese organisations in the manufacturing and services sectors, it has been found that 'TQM was part of the daily language and activities in the organisations'. The whole environment inside the organisations would reflect the embracing of quality. In other words, 'TQM was integrated into the firms' management practices and operations [as with EFQM EM]'.

Not only does policy and strategy need to be well planned and developed but it also needs an inclusive approach if it is to be deployed at all levels within a construction organisation. Oakland et al. (2002) suggested using scorecards as a tool to be able to measure, review and update the policies during all stages, which will result in achieving a consistent approach to measuring progress towards the set organisational objectives.

People

Excellent organisations manage, develop and release the full potential of their people at an individual, team-based and organisational level. They promote fairness and equality and involve and empower their people. They care for, communicate, reward and recognise, in a way that motivates staff and builds commitment to using their skills and knowledge for the benefit of the organisation.

SUB-CRITERIA

3a. People resources are planned, managed and improved.
3b. People's knowledge and competencies are identified, developed and sustained.
3c. People are involved and empowered.
3d. People within the organisation have a dialogue.
3e. People are rewarded, recognised and cared for.

Figure 4.6 shows the linkages to the People criterion.

The People criterion could assist in obtaining optimum results, if there existed in the company a well-developed human resource policy, strategies, and plans which included and encouraged training, empowerment, communication and teamwork.

'The concept of internal and external customers/suppliers forms the core of a total quality approach and the basis of Business Excellence' (Oakland et al. 2002). Many senior managers underestimate the impact of people in obtaining an organisation's aims and objectives incorporated within their mission and vision statements.

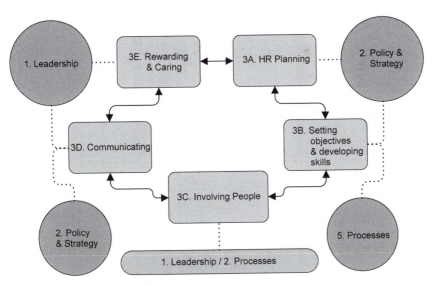

Figure 4.6 Linkages within the People criterion

Moreover, Oakland et al. (2002) suggested that the essence of the EFQM Excellence Model is its mechanism, dynamics, and recognition of the importance of the role of employees, and that 'Processes are the means which a company or organisation harnesses and releases the talents of its people to produce results-performance'.

Partnerships and resources

Excellent organisations plan and manage external partnerships, suppliers and internal resources in order to support policy and strategy and the effective operation of processes. During planning and whilst managing partnerships and resources they balance the current and future needs of the organisation, the community and the environment.

SUB-CRITERIA

- 4a. External partnerships are managed.
- 4b. Finances are managed.
- 4c. Buildings, equipment and materials are managed.
- 4d. Technology is managed.
- 4e. Information and knowledge are managed.

Figure 4.7 shows the linkages to Partnerships and Resources.

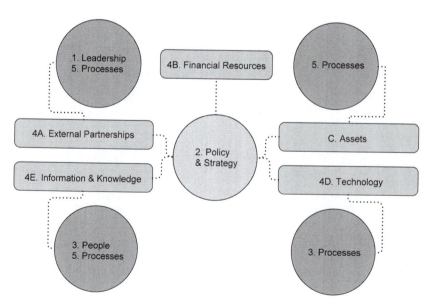

Figure 4.7 Linkages within the Partnerships and Resources criterion

There is a focus in the UK on the attainment of 'best value'; the fundamental principles of best value have been outlined by Donnelly (1999):

- accountability;
- transparency;
- continuous improvement;
- ownership;
- to be delivered through sound governance, long-term planning and budgeting control;
- application of performance management (including a top quality services provision).

In addition, the principles of best value must provide the driving force for developing a corporate and service quality strategy. UK Government information on the attainment of 'best value' advocates the utilisation of partnering initiatives and of course the efficient and effective use of corporate resources.

Moreover, value for money is an imperative requirement in all construction organisations.

Partnering

Partnering may be defined as a long-term commitment between two or more construction-related organisations for the purpose of achieving specific business objectives by maximising the efficiency and effectiveness of each participant's resources. This requires changing traditional relationships to a shared culture without regard to organisational boundaries. The relationship should be based on trust, dedication to common goals, and an understanding of each other's individual expectations and values. Expressed benefits of engaging in partnering include improved efficiency and cost-effectiveness, increased opportunity for innovation, and the continuous improvement of quality products and services.

In addition partnering can be split into two main types:

- Project partnering; this is a partnering arrangement on a single project in which, at the end of the project, the partnering relationship is terminated.
- Strategic partnering is a partnership arranged on a long-term basis, in order to gain long-term synergistic advantages.

Experience has shown that strategic partnering provides more organisational benefits when compared to project partnering; this is because it allows time for continuous improvement and relationships to be developed over a sustained period of time.

Processes

Excellent organisations design, manage and improve processes in order to fully satisfy, and generate increasing value for, customers and other stakeholders.

SUB-CRITERIA

5a. Processes are systematically designed and managed.
5b. Processes are improved, as needed, using innovation in order to fully satisfy and generate increasing value for customers and other stakeholders.
5c. Products and services are designed and developed based on customer needs and expectations.
5d. Products and services are produced, delivered and serviced.
5e. Customer relationships are managed and enhanced.

Figure 4.8 depicts the linkages with Processes.

To be able to incorporate the five sub-criteria, which would lead to the enhancement of processes for construction organisations, it is crucial to apply three key principles. These are project management, culture

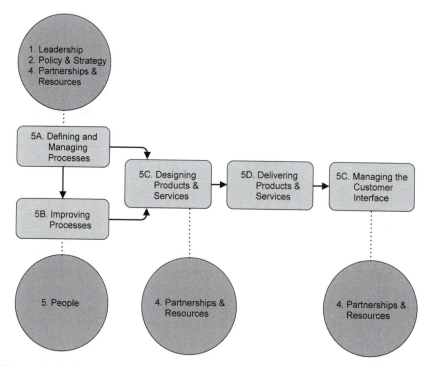

Figure 4.8 Linkages within the Processes criterion

change and benchmarking, encapsulated within the concepts of a learning organisation.

Process management

Orsini (2000) outlined the following steps for the achievement of process improvement:

- provide a clear definition of the processes under consideration;
- identify all the interactions of the processes under consideration with all other processes;
- produce a specification of the critical characteristics of the process under consideration;
- produce a means of measuring the critical characteristics of the process under consideration;
- collect sufficient and reliable data and determine whether the process under consideration is unstable or stable;
- establish the key performance indicators in order to be able to take advantage of any improvement opportunities that may develop;
- be prepared to make changes [this can best be accomplished by using the RADAR approach].

'Failure to meet requirements in any part of a quality chain has a way of multiplying and failure in one part of the system creates problems elsewhere [thus the integrative nature of processes must be a prime consideration for companies]' (Orsini 2000). This concept is so significant that if senior management and employees can grasp it, it will ease the transformation of a cultural change and would thus lead to the accomplishment of a 'continual improvement culture'. Since this approach is a continuous process, Oakland et al. (2002) noted that in public-sector organisations the process criterion consists of four elements:

- To provide direction and improvement.
- To satisfy customer needs.
- To manage and ensure the capability of the organisation.
- To engage in the measurement function and improve organisational performance.

Processes integrated into one system

One of the fundamental problems associated with the deployment of any TQM/EFQM EM activity is the failure to appreciate the integrative nature of the system. Orsini (2000) pointed out that if all employees worked as hard as they possibly could, it would not be sufficient to solve the problematic issues of a poorly designed organisational system.

In addition, Orsini (2000) argued that the condition of such an organisation will lead to poor communication and the ineffective implementation of corporate objectives. He summarised the key characteristics of such poor performance as:

- inconsistencies in organisational procedures;
- failure to properly manage people;
- destructive internal competition;
- sub-optimisation between groups, within a department, within a division, within a company, between a company and its suppliers, between a company and its customers;
- a failure to think or plan ahead;
- trying to engage in retrospective corrective actions 'after the fact', exhibited through audits, inspections and untimely feedback.

Table 4.2 provides a summary of key deployment issues along with the resulting advantages of deployment for construction organisations.

Culture

Culture is related to social anthropology, which is an aspect of the study of the 'shared meanings and values held by groups in society that give significance to their actions' (McKenna 2000). However, in construction organisations it is usually referred to as corporate culture, which comprises 'behaviour, actions, and the values that people in an enterprise are expected

Table 4.2 EFQM Excellence Model deployment advantages for construction organisations

Key Deployment Issues	Resulting Benefits
Process improvements	A clear understanding of how to deliver value to clients and hence gain a sustainable competitive advantage via operations.
Attaining an organisation's objectives	Enabling the mission and vision statements to be accomplished by building on the strengths/ distinctive competence of the company.
Benchmarking key performance indicators (KPIs)	Ability to gauge what the organisation is achieving in relation to its planned performance (Plan, Do, Check, Act). And engage in continuous organisational improvement.
Development of clear, concise action plans resulting in a focused policy and strategy	Clarity and unity of purpose so the organisation's people can excel and continuously improve.
Integration of improvement initiatives into normal operational activities	Interrelated activities systematically managed with a holistic approach to decision making resulting in a learning and improving culture.

(Note: issues are not mutually exclusive)

to follow.' While other researchers defined organisational culture as 'a pattern of basic assumptions, invented, discovered, or developed by a given group as it learns to cope with its problems of external adoption and integral integration, that has worked well enough to be considered valid and, therefore, is to be taught to new members as the correct way to perceive, think, and feel in relation to those problems' (McKenna 2000).

Dooley (1998) addressed an important aspect of cultural change and the correct approaches to attaining such a sea change. There are two main approaches to be considered when deploying a change strategy, the first being a focus on an 'individual's cognitive processes'. The second approach relates to 'their actions and interactions within the firm'. These two approaches are referred to as Thoughts and Action.

The first approach is referred to as the 'thoughts approach', where individual behaviour is attributed to their knowledge, attitudes and beliefs. Change can occur through an organisational training strategy that is supported by consistent, rational and coherent application. The second approach to change as noted is 'actions'; it is based on the assumption that an employee's knowledge, attitudes and behaviour are shaped by employee involvement (empowerment).

Changing the culture

Changing an organisation's culture is a vital component in the successful implementation of TQM/EFQM EM. Organisational change processes must be managed; first, the change of culture must be part of an explicit and comprehensive plan for improvement. Second, senior and middle management must be seen to be the driving force; in other words 'managers must learn to lead change in an overt manner'. Third, before starting to develop the plans for change, the views of managers and employees should be obtained and evaluated, the results of questionnaires, group meetings, etc., can provide the means for the information-gathering process. Fourth, cultural change should be introduced as an ongoing process rather than a requirement of the introduction of TQM/EFQM EM. Fifth, to facilitate cultural change construction organisations will benefit from the utilisation of management tools and techniques, used effectively and with purpose. TQM/EFQM EM application does not negate the use of management tools such as statistical process control (Dale 1999).

In addition, Dale (1999) emphasised that employees' roles are a significant component of cultural change. Since employees are the real assets of an organisation, he proposed that the value of this asset (employees) would increase or decrease according to the way employees are treated. Thus, Dale proposed that for a cultural change to be successful, the following should be considered: the fact that people have different backgrounds, ages, skills, abilities, levels of enthusiasm, levels of competency, levels of flexibility and ability to accept change.

Benchmarking

One of the key strengths of the EFQM EM is the self-assessment methodology which forms a strong basis for benchmarking performance with other organisations. Benchmarking can be defined as 'a positive, proactive process to change operations in a structured fashion [as part of a learning organisation culture] to achieve superior performance' (Ball et al. 2000).

Ball et al. (2000) attributed the success of benchmarking in the private sector to being utilised as a method for searching for new ideas and practices for adoption in a company and assisting in the attainment of competitive advantage. Benchmarking is used as a management tool for the improvement of organisational performance by engaging in organisational learning.

Customer results

Excellent organisations comprehensively measure and achieve outstanding results with respect to their customers.

A) Perception measures: These measures are of the customer's perceptions of the organisation (obtained, for example, from customer surveys, focus groups, vendor ratings, compliments and complaints feedback).

Some suggested measures of customer results are:

- Organisation's image: accessibility, communication and responsiveness.
- Products and services: quality, value, reliability, design innovation, relevance of the products or service, delivery and environmental profile.
- Product or service and aftercare: capabilities and behaviour of employees, advice and support; handling complaints, response time.

(EFQM 2000b)

B) Performance indicators: These are the internal measures that can be used by an organisation in order to monitor, understand, predict and improve the performance of the organisation; they may also be able to predict the perception of its external customers.

Some suggested measures based on EFQM publications (2000b) are:

- Organisation's image: press coverage – it is important for people to be aware of any projects or activities that support economic growth or contribute towards a 'social audit'.
- Products and services: Competitiveness/value for money; defect, error and rejection rates. Performance against customer-based objectives: complaints; product life cycle; innovation in design; time taken to deliver to society.

- Product or service support and aftercare: Handling of complaints and response rates.
- Loyalty: Duration of relationship; effective recommendations; numbers of complaints and compliments and commendations.

People results: Excellent organisations comprehensively measure and achieve outstanding results with respect to their people.

A) Perception measures: These measures are of the people's perception of the organisation (obtained, for example, from surveys, focus groups, interviews, structured staff appraisals).

Examples include:

- Motivation: career developments; communication; empowerment; equal opportunities; leadership; opportunity to learn and achieve; recognition; target setting and appraisal; the organisation's values, mission, vision, policy and strategy; training and development.
- Satisfaction: the organisation's administration; employment conditions facilities and services; health and safety conditions; job security; pay and benefits; peer relationships; the management of change; the organisation's environmental policy and impact; the organisation's role in the community and society; working environment.

B) Performance indicators: These measures are internal ones and are used by organisations in order to monitor, understand, predict and improve the performance of the organisation's people and to predict their perceptions.

Examples of such measures would incorporate:

- Achievements: competency requirements versus competencies available; productivity; success rates of training and development to meet set corporate objectives; external awards and recognition.
- Motivation and involvement: involvement in improvement teams; involvement in suggestion schemes; levels of training and developments; measurable benefits of team work; recognition of individuals and teams; response rates to stakeholder surveys.
- Employee satisfaction: absenteeism and sickness levels; accident levels; recruitment trends; staff turnover and loyalty; use of organisation-provided facilities and benefits.
- Services provided to the organisation's employees; accuracy of personnel administration; communication effectiveness; speed of responses to enquiries; training evaluation and feedback.

Society results: Excellent organisations comprehensively measure and achieve outstanding results with respect to society.

A) Perception measures: These measures are of the society's perception of the organisation (obtained, for example, from surveys, reports, press articles, public meetings, public representatives, governmental authorities). Some of the measures contained in the guidance for perception measures may be applicable to performance indicators and vice versa.

For example an organisation could employ the following feedback mechanisms:

- Image: responsiveness to contacts; as an employer; as a responsible member of the community.
- Performance: disclosures of information relevant to the community; equal opportunity practices; impact on local, regional, national and global economies; relationships with relevant authorities; ethical behaviour.
- Involvement in the communities where it operates: involvement in education and training and community bodies in relevant activities; support for medical and welfare provision; support for sport and leisure; reduction and prevention of any nuisance and harm originating from its operations and/or throughout the life cycle of its products and services; health risks and accidents; hazards (safety); pollution and toxins emission; environmental performance evaluation/life cycle evaluation.
- Reporting on activities to assist in the preservation and sustainability of resources: reduction and elimination of waste and packaging; substitution of raw materials or other inputs; usage of utilities, e.g. gases, water, electricity; recycling.

B) Performance indicators: These measures are the internal ones utilised by companies in order to monitor, understand, predict and improve perceptions of it by society.

- Handling changes in employment levels.
- Dealing with authorities on issues such as: certification; clearances import/export; planning; product release.
- Accolades and awards received: exchange of information in relation to social responsibility good practices, auditing and social reporting.

These are several examples of measures based on EFQM publications (2000b):

- Key performance results: Excellent organisations comprehensively measure and achieve outstanding results with respect to the key elements of their corporate policy and strategy.

A) Key performance outcomes: These measures are key results defined by the organisation and agreed in their policy and strategies and may include:

- Financial outcomes: meeting of set budgets; audited accounts including income, grants and expenditures.

- Non-financial outcomes: time to market; success rates as defined by the corporate mission and vision statements; compliance with legislation and codes of practice; results of statutory audits and inspections; process performance information.

B) Key performance indicators: These measures are the operational ones used in order to monitor and understand the processes, and predict and improve the organisation's performance outcomes; they could incorporate:

- Financial: cash flow and maintenance costs.
- Non-financial: processes (performance; assessments; innovations, cycle times); external resources including partnerships (supplier performance; supplier price; number and value added of partnerships; number and value added of joint improvements with partners); buildings, equipment and materials; (defect rates; inventory turnover; utilisation); technology (innovation rate; value of intellectual property; patents; royalties); information and knowledge (accessibility; integrity; value of intellectual capital).

Difficulties in interpreting results

Dooley (1998) discussed some of the typical problems associated with interpreting the results of TQM models and pointed out that a common problem of interpretation is that results are interpreted by people. Therefore, results will be influenced by the perceptions of employees. To comprehend employees' perceptions, managers must be aware of the principles of organisational behaviour studies. However, within service and manufacturing companies many managers and engineers fail to recognise the importance of organisational behaviour. There can exist a tendency to concentrate on the technical aspect of a company, probably because managers are personally more comfortable dealing with these issues.

Another common problem is that of employees not being prepared to 'raise red flags'. The reason for such behaviour in employees can be attributed to the fact that they are insecure. This would demonstrate a culture that is not conducive to TQM/EFQM EM application; it would be indicative of a culture where failure brings sanctions rather than a true desire to investigate and eradicate the root cause of the problem and hence improve.

A further issue could be that of the time lag between cause-and-effect measurement. Because of the dynamic nature of organisations and since strategies are chosen based on a perceived organisational condition at a specific point in time, changes may occur while implementing quality initiatives. This may generate unreliable results if the changes are not taken into account. In order to achieve consistency in measuring results within companies, a specific methodology should be established and Capon et al. (1995) suggested the use of an approach known as 'process capability'. It is a measure of 'errors measured each week and compared with defined standards and it can then be expressed as a percentage of all transactions'.

Performance measurement

Capon et al. (1995) addressed an important aspect related to organisational results measurement; this they did in the sense that TQM results are dependent upon what stakeholder's perceptions and satisfaction levels are. They emphasised that a TQM programme that focuses purely on technical objectives, such as the control of incurred overhead costs, would fail. This is because they should be related to other corporate activities and objectives; in other words, the strategies are not mutually exclusive.

The best approach for measuring the success of TQM/EFQM EM is to consider the internal and external factors which have an impact on the host organisation (Capon et al. 1995).

Moreover, the main objective of performance measurement is to make sure that all sub-elements of the whole organisational system are working effectively and efficiently to achieve the required results. Thus Castka et al. (2003) noted that for an organisation to be able to measure its key performance results effectively and efficiently, it needs to consider adopting a holistic approach to performance measurement that is divided into three levels:

- Organisational level: organisational relationship to its markets; the variables that affect performance at this level are organisational strategies, goals, objectives, organisational structure and deployment of its resources.
- Process level: focuses on work flow within the organisation; the process level is related to the output of the organisation; performance must meet the needs of the customer.
- Job/performer level: processes are managed by individuals; typical variables include hiring and promotion, job responsibilities and standards, feedback, rewards, and training.

In fact, Castka et al. (2003) proposed that there is a need to establish a performance measurement system that would include and manage the following activities:

- the identification and prioritisation of desired results;
- establishing the means to measure progress towards prioritised desired results;
- setting [valid] standards for assessing how satisfactorily results have been achieved;
- tracking and measuring progress towards [the attainment of] results;
- exchanging ongoing feedback among those participants working to achieve results;
- periodically reviewing progress;
- reinforcing activities that achieve results; and
- intervening to improve progress where [and when] needed.

Performance indicators

For a performance measurement system to work optimally a company must establish valid performance indicators (PIs) and examples could include:

- **Financial versus non-financial**: the traditional use of financial PIs alone is no longer a valid approach; they are not sufficient to determine the organisation's holistic health and other types of indicators are necessary for e.g. the current age profile of its existing staff base.
- **Global versus local**: global PIs are usually the preserve of senior management and used within the corporate planning process and local PIs are usually utilised by middle managers and normally relate to operational activities.
- **Internal versus external**: internal PIs are used to monitor the performance of an organisation on aspects that are relevant to its internal functioning, while external PIs are introduced to evaluate the performance of the organisation as experienced by customers. Or to evaluate the performance of the organisation as experienced by customers or to evaluate the performance of suppliers.
- **Organisational hierarchy**: performance indicators can be placed in a hierarchical order; this is to say that PIs should relate to specific organisational levels, it is really very similar to the use of 'management by exception'.

Key performance results

Performance measurement is 'the process of assessing progress towards achieving predetermined goals'. Utilising such a process will lead to the gaining of insight into, and being able to make valid judgements about, the effectiveness and efficiency of their programmes, processes and people (Agere and Jorm 2000).

The EFQM EM divides the key performance results criterion into two sub elements:

1. Key performance outcomes: the measures are defined by the organisation and agreed in policy and strategies and can be categorised as:

   (a) Financial outcomes
   (b) Non-financial outcomes

2. Key performance indication: these measures are the ones used in order to monitor and understand the process and predict and improve.

It is important to note that 'the involvement of all managers and employees in determining the type of measurement required is good practice, in order to make them committed to its success and for them to own the process' (Agere and Jorm 2000).

The establishment of performance targets has proved to be a key factor in the improvement of the quality of service and value received by customers. However, there are two aspects to consider; first, achieving effectiveness which is the extent to which customer requirements are met. Second, attaining efficiency which is the measure of how an organisation's resources have been used in providing said customer satisfaction. In addition, it is critical for the success of a performance measurement system to include both financial and non-financial measures, and more importantly to consider all stakeholders and their requirements (McAdam et al. 2002).

Knowledge measurement

Many organisations do not consider employees' knowledge as a scarce resource. It is not just what and how, many individuals learn, but how effectively they can transfer their knowledge throughout the construction organisation and how effectively it can be applied (Agere and Jorm 2000).

Agere and Jorm (2000) have emphasised the importance of developing an effective appraisal system and they further outlined some of the key benefits of applying such a system as follows:

- to ensure that the performance of staff is measured against the set [organisational] objectives;
- to identify the strengths and weaknesses of staff and to prepare a programme that will both build on their strengths and address their weaknesses in line with set objectives;
- to develop a mutually agreed action plan for the development of staff in line with set corporate objectives;
- to create an appropriate climate for two-way discussion to the mutual advantage of both staff and organisation [the application of a 360-degree appraisal system would ensure a reciprocal feedback loop].

RADAR

At the heart of the EFQM excellence model a specific logic exists and this is known as RADAR. The RADAR logic consists of the following four elements: Results, Approach, Deployment, Assessment and Review.

Table 4.3 provides a succinct overview of the RADAR concept.

The key concepts built into RADAR are that the learner experience is critical to self-assessment, with an emphasis on collecting valid and reliable evidence, that the primary purpose of self-assessment and development action plans is self-improvement. All assessments are required to be deployed with rigour, irrespective of organisational size. The host organisation should always try and triangulate its collected evidence base; reflection upon results is vital in order to engage in triple-loop learning.

Table 4.3 Succinct overview of the RADAR concept

Results
This covers what an organisation achieves. In an excellent construction organisation the results will show positive trends and/or sustained good performance, which will compare well with others and will have been caused by the adopted approaches. Additionally, the scope of the results will address the relevant areas.

Approach
This covers what a construction organisation plans to do and the reasons for it. In an excellent construction organisation the approach will be sound – having a clear rationale, well defined and developed processes and a clear focus on stakeholder needs – and will be integrated, supporting policy and strategy and linked to other approaches where appropriate.

Deployment
This covers what a construction organisation does in order to deploy the approaches. In an excellent construction organisation the approaches will be implemented in relevant areas, in an appropriate systematic way.

Assessment and Review
This covers what a construction organisation does to assess and review both the approaches and the deployment of the adopted approaches. In an excellent construction organisation the approach, and the deployment of it, will be subject to regular measurement, learning activities will be undertaken, and the output from both will be used to identify, prioritise, plan and implement improvement activities.

The application of the RADAR philosophy will assist in driving business improvement through utilising the model. The logic purports that a construction organisation needs to:

- Determine the **Results** it is aiming for as part of its policy- and strategy-making process. These results cover the performance of the organisation, both financially and operationally, and the perceptions of its stakeholders.
- Plan and develop an integrated set of sound **Approaches** to deliver the required results both now and in the future.
- **Deploy** the approaches, in a systematic way, to ensure full implementation.
- **Assess and Review** these approaches based on monitoring and analysis of the results achieved and ongoing learning activities. Finally, identify, prioritise, plan and implement improvements where required.

When using the model within a construction organisation the Approach, Deployment, Assessment and Review elements of the RADAR logic should be addressed for each enabler criterion part and the Results element should be addressed for each results criterion part.

Figure 4.9 Deming's dynamic control loop cycle

EFQM's RADAR model mechanism is related to Deming's continuous improvement philosophy of Plan – Do – Think – Act (see *Figure 4.9*). More importantly, the process is driven by self-assessment, which is not only a means for measuring continuous improvement, but also an excellent opportunity to integrate Total Quality Management into normal operations (EFQM 2000b).

Finn and Porter (1994) noted the foremost reason for using the EFQM Excellence Model was that it offered a 'framework for exploring the link between organised activities and results and for driving continuous improvement'. Further, Westlund (2001) identified that the 'EFQM Excellence Model has clearly become the most applied model in Europe for total quality management (TQM)'.

Osseo-Asare Jr and Longbottom (2002) proposed that the nine criteria represent decisive success factors and encompass TQM principles. The application of the EFQM Excellence Model within a construction organisation can offer 'a framework for exploring the link between organisational activities and a drive for continuous improvement'. Curry (1999) summarised the uniqueness of EFQM EM by stating that 'this model provides a tangible framework for assessing excellence in an organisation and for making step improvements in operations. It helps to bring greater cohesion to the different organisational activities'.

In addition, Coleman and Douglas (2001) argued that the EFQM EM 'defines and describes TQM in a way that can be more easily understood by senior management'. This would allow them to accept ownership of any changes required and be able to drive their organisations towards excellence. It would also provide a tangible pathway to TQM, with clearly defined requirements. And more importantly, they suggested that any lack of clear requirements for achieving TQM are not to be found in the EFQM Excellence Model.

Furthermore, there is another significant aspect that makes the choice of the EFQM Excellence Model much more appropriate than Baldrige's National Quality Award (MBNQA) and others. The EFQM Excellence Model was developed and based on lessons learned from previous experiences and other TQM models such as MBNQA. Organisations make use of self-assessment models as a tool to provide a path for 'What they should do' (Ho 1999).

Osseo-Asare Jr. and Longbottom (2002) outlined the main characteristics of EFQM Excellence Model as follows:

- It provides a holistic way of managing a business enterprise, which will lead to its long-term success.
- The model is a diagnostic tool for self-assessment of the current health of an organisation. Self-assessment will provide the ability to balance an organisation's priorities, allocate resources and generate realistic business plans.

Pitt (1999) pointed out the reasons for choosing the EFQM Excellence Model over other models. First, the EFQM EM provides a broader approach to quality assurance and continuous improvement, in comparison to ISO 9001:2008. Second, EFQM EM has been designed by European organisations; it can be said that the EFQM EM has a certain 'European Flavour', so it is easy to benchmark against other construction organisations across Europe. Third, the EFQM EM was developed and improved based upon other TQM models such as the Malcolm Baldrige Model. Fourth, the model is capable of integrating other quality initiatives such as ISO 9001:2008, Investors in People, and the Charter Mark.

Furthermore, Ho (1999) suggested that an important reason behind choosing a particular TQM framework over another was the geographic location of the firm. There are three major quality awards: Japan's Deming Prize, the USA's Malcolm Baldrige National Quality Award (MBNQA) and the European Quality Award (EQA). Therefore, organisations located in

Europe, or who have a strong European presence, would find the EFQM EM most appropriate. Choosing a model according to the location will allow for benchmarking with other organisations working within the same operational environment (Ho 1999).

Question 1 for the reader

The European Foundation for Quality Management (EFQM) has stated that the functions of their Excellence Model may be split into four components. Identify these four component parts. The answer is provided at the end of the book.

Self-assessment processes

The European Foundation for Quality Management provides a useful definition of self-assessment for quality management (EFQM 2000b). Self-assessment is

> a comprehensive, systematic and regular review of an organisation's activities and results against the EFQM Excellence Model. The Self-Assessment process allows an organisation [suitable for construction organisations] to discern clearly its strengths and areas in which improvements can be made. Following this process of evaluation, improvement plans are launched, which are monitored for progress.

One of the unique aspects of the EFQM EM is that it enables a reverse direction to self assessment to be adopted. 'This means adopting a diagnostic sequence, which starts from results (or symptoms)' (Conti 1997).

The first phase of self-assessment for construction firms consists of analysing results in order to establish performance gaps. Second, scores may be allocated in order to pinpoint areas to be addressed, thus engaging in continuous improvement processes.

Dale (1999) argued that self-assessment is an effective tool for achieving continuous improvement. He postulated that before embarking on self-assessment, both management and employees need to have some knowledge of self-assessment in order that they can understand the questions related to the self-assessment process.

Dale (1999) described the mechanism of self-assessment as follows: 'it uses one of the models underpinning an award to pinpoint improvement opportunities and to identify new ways in which to encourage the organisation down the road of business excellence'.

> The concept of organisational learning involves teaching an organisation to use a scientific method (self-assessment), to create and utilise specific knowledge, and to change its performance measurement system.
>
> (Hendricks and Singhal 2001)

The EFQM EM, with its 'RADAR' application as an integral part of the model, does enable a positive change in organisational performance to be achieved, and thus continued improvement is an attainable organisational objective.

Different approaches to self-assessment

There are different approaches to conducting a self–assessment process and these are now outlined; construction firms need to fully consider which approach is most suitable for them.

The award simulation approach

Although this is potentially the most time consuming and resource intensive of all the approaches described, it is very comprehensive. It will provide one of the most accurate scoring profiles, allowing for legitimate comparisons with the scoring profiles of award applicants.

AWARD SIMULATION ADVANTAGES

- It produces a list of strengths and areas for improvement developed by a team of trained assessors which can be used to drive improvement actions.
- The process of writing the information down provides a powerful and concise way of reflecting the culture and performance of the construction company. As it is a written report it can be referred to repeatedly.
- Once the first report has been completed, subsequent reports are relatively easy to complete with a high degree of accuracy and consistency.
- It provides an excellent opportunity for involvement and communication during the data-gathering process. When completed it also provides an excellent communications document to be shared by the people within the company, its customers, suppliers and others with an interest in the organisation. Some construction organisations use the report as part of their marketing strategy.
- The site visit and presentation from the assessor team are important value-adding steps since they provide an opportunity to check deployment issues, and for the assessors to explain, in greater detail, the rationale behind their comments in the feedback report.
- The process provides a learning opportunity prior to application for the Quality Award for Excellence.
- It provides an easy way for units within a construction company to compare processes and results and identify examples of best practice that may be shared/disseminated throughout the organisation.

DISADVANTAGES

- There is a temptation for a management team to be less involved by taking the opportunity to delegate most of the work.

- It can be seen as an exercise in creative writing, covering up the real issues.
- For those construction organisations in the early days of their journey to Excellence, this approach may be too ambitious as a first attempt at self-assessment. This is because it is very resource intensive.

The ASSESS questionnaire approach

This is a comprehensive approach to self-assessment incorporating simple and easy-to-use questions and providing a focus on actionable data to aid improvement planning and benchmarking.

ASSESS QUESTIONNAIRE ADVANTAGES

- Questions are developed from the model and so provide full coverage of the noted criteria for construction enterprises.
- A quick and easy entry point to self-assessment through Rapid Scoring, a more searching self-assessment through Team Scores, and an independently validated self-assessment.
- Scoring has been calibrated against the full award process and the model.
- The approach is software or paper based, or can be run in tandem.
- Each question is supported by hints and tips which provide more detailed insight into the questions.
- Is a very powerful learning approach to self-assessment for the individual and company.
- The software can capture narrative as well as the quantitative scores.
- The process can be undertaken individually or as a group.
- Individual assessments can be merged through the use of a powerful team facility in the software, providing a focus for those areas that require consensus.
- Software provides export facilities to other reports and presentations.
- Has an excellent graphics facility.
- Output can be entered into a database for benchmarking purposes that will be compatible with the national benchmarking network.

DISADVANTAGES

- Needs a certain degree of software but can be paper based.
- Can be very searching of individuals and organisations as the questions are quite specific.

The pro-forma approach

Although the data-gathering part of this process might be as long as the award simulation approaches, the task of preparing the pro-forma, one page per part criterion, is easier and less time consuming than drafting a full award-style report.

PRO-FORMA ADVANTAGES

- Provides a list of strengths and areas for development for driving improvement actions for construction enterprises.
- Allows people to document the evidence upon which strengths, areas for improvement and scores are based.
- Scoring profiles can be derived which, in terms of accuracy, lie closer to the award simulation approach rather than, for instance, the matrix chart approach.

PRO-FORMA DISADVANTAGES

- The collection of pro-formas may not inform of the full story of the organisation; it represents a summary of the position.

The workshop approach

In terms of resources required, this approach does not take as long as the award simulation process but on average is likely to take longer than either the matrix chart or questionnaire approaches.

WORKSHOP ADVANTAGES

- It is probably the best way to get a management team to understand and commit to the model and its deployment.
- Discussion and agreement by the management team on the strengths and areas for improvement helps to build a common and comprehensive view on the current state of the construction organisation. This leads to ownership by the management team of the output and facilitates their prioritisation and agreement of action plans.
- Provides a building opportunity for the management team.
- An agreed list of strengths and areas for improvement is produced which will drive improvement actions for all.

DISADVANTAGES

- A less robust and rigorous process than the award simulation approach.
- Can be a high-risk approach and needs excellent preparation and facilitation to ensure the management team are fully prepared and comfortable with the process. Ground rules for behaviour during the workshop should be agreed and understood beforehand.

The matrix chart approach

This approach is less resource intensive and quicker to use than the award simulation approach provided an existing matrix chart is used. However, the resource and time requirements will increase considerably if a construction organisation chooses to create its own matrix chart. It is particularly suited for use by small teams.

MATRIX ADVANTAGES

- It is simple to use, as basic awareness training is sufficient to get things started.
- Can be used to involve everyone in self-assessment.
- Provides a practical way of understanding the criteria.
- Provides a means for teams to assess their progress quickly and easily and progress can be readily displayed. Gaps can also be clearly demonstrated, giving an indication of what to do next.
- Good for facilitating team discussions and team building.
- Involving the management team in developing its own matrix chart can be a powerful process and it forces them to discuss, reach consensus, articulate their collective vision, and describe the steps towards achieving it in all nine criteria areas.

MATRIX DISADVANTAGES

- It does not provide an 'award standard' self-assessment; lists of strengths and areas for improvement are not produced.
- It does not allow for comparisons with award applicants.
- There is not necessarily a direct link between the steps in the matrix chart and the criterion parts of the model.

The peer involvement approach

Similar to award simulation in terms of providing a comprehensive approach to self-assessment, with the associated time and resource implications.

PEER INVOLVEMENT ADVANTAGES

- Less prescriptive than the award simulation approach. The unit undergoing self-assessment does not have to produce a full report; the submission can be in any suitable form.
- Provides the opportunity for the involvement not only of people within the unit but also their colleagues from other parts of the construction organisation. This leads to a high degree of cross-functional learning for the assessors and the company.
- Provides a comprehensive list of strengths and areas for improvement for driving improvement actions, again for all.

PEER INVOLVEMENT DISADVANTAGES

- It can require the use of more resources than some of the other approaches.

- The degree to which units would volunteer for the exercise and be prepared to share information may limit the value of the exercise.

The simple questionnaire approach

This approach is one of the least resource intensive and can be completed very quickly, provided an existing and proven questionnaire is used. It is an excellent approach for gathering information on the perceptions of people within the construction firm.

ADVANTAGES

- It is simple to use and some basic awareness training is sufficient to commence activities.
- It can readily involve many people within the company.
- Presentations of outcomes are less problematic.
- Easy to compute and understand the numerical results.
- The questions asked can be customised to suit the construction company.
- It does provide a good introduction to self-assessment.
- Enables the organisation to receive feedback which can be segmented by function and level.
- Can be used to facilitate group discussions between teams on the opportunities for improvement within their units and as a total improvement process.

DISADVANTAGES

- Excessive use of questionnaires in any organisation may result in a low return. What response rate is a valid return?
- Not everyone in the company may understand the meaning of the questions.
- Wide circulation can raise expectations amongst the people within an organisation and the use of this approach will need careful positioning.
- Questionnaires tell you what people think, not why they think that way.
- A list of strengths and areas for improvement is not usually generated.
- Does not allow for comparison with scoring profiles of award applicants.
- Accuracy and relevance depend upon the quality of questions asked.

(British Quality Foundation 1998)

In brief, Dale (1999) advocated that the process of self-assessment will constitute three main stages:

1. The gathering of data for each criterion related to the model.
2. Conducting a valid assessment of the data gathered.
3. Developing appropriate plans and actions arising from the assessment and monitoring the progress and effectiveness of the plan of action.

Dale (1999) further outlined the key issues to be considered by organisations embarking on self-assessment:

- Ensure that senior management are committed to the self-assessment process and are prepared to use the results to develop improvement plans [not an easy task].
- Arrange for everyone involved in the process who requires some training to be trained.
- Communicate to all the rationale for engaging in the self-assessment process.
- Plan the means for collecting the data.
- Decide on the team and allocate roles and responsibilities for each criteria of the model.
- Develop a valid data-collection methodology and identify data sources.
- Agree an activity schedule and manage it as a project.
- Decide the best way for organising the data which has been collected.
- Present the data, reach agreement on strengths and areas of improvement and agree the scores for the criteria.
- Prioritise the improvement and develop an appropriate action plan.
- Conduct regular reviews of progress against the plan.
- Repeat the self-assessment process as appropriate.

The uniqueness of the self-assessment processes lies in providing real evidence that can be utilised in the form of a trend analysis, thus enabling a construction organisation's momentum towards TQM/EFQM EM to be monitored and enhanced, encapsulated within the RADAR concept.

Benefits of using EFQM/self-assessment

Castka et al. (2003) note the benefits of using EFQM/self-assessment as:

1. Providing the opportunity to take a broader view on how the measured activity is impacting on the various business operations.
2. Measuring performance of processes, enablers and their relationship with organisational results.
3. Self-assessment conducted both internally and externally to the organisation.
4. Providing an opportunity to benchmark and compare like for like, or
5. Measurement for providing improvement rather than for hard quality control; and
6. Self-assessment is also an important communication and planning tool:

   6.1 The results of self-assessment provide a growing common language through which organisations, or parts of organisations, can compare their performances.

6.2 The outputs of self-assessment are used for strategic management and action planning, or as a basis for an improvement project.

6.3 New business values: leadership, people, process management, the use of information within the organisation and the way customer relationships are managed.

Underpinning the EFQM Excellence Model are the principles of knowing where a construction organisation is, where it wants to go and how it can get there. The model links self-assessment to informed planning and to implementation, through 'a framework of key processes'. Self-assessment can be seen as a catalyst for driving business improvement and hence achieving business goals.

Using a combination of methods

Self-assessment methods based on workshops and pro-formas can involve relatively few people within an institution, although this is very much dependant on what the whole assessment process looks like. Questionnaires can provide extra data from a much wider base and thereby support either the pro-forma or workshop method.

A combination of pro-forma and workshops is useful. Pro-formas enable the gathering of a lot of detail and this – when carefully collated and presented – is an excellent basis for workshops, where the issues and supporting detail may be fully explored (EFQM 1999a, 1999b).

Deploying the European Foundation for Quality Management Excellence Model

Having recognised that corporate excellence is measured by an organisation's ability to both achieve and sustain outstanding results for its stakeholders, the enhanced version of the EFQM Excellence Model was developed. The fundamental advantages of the Excellence Model included:

- Increased cost effectiveness; results orientation; customer focus; partnership; knowledge management; performance and learning.
  (European Foundation for Quality Management 1999a)

The Excellence Model was designed to be:

- Simple [easy to understand and use]; holistic [in covering all aspects of an organisation's activities and results, yet not being unduly prescriptive]; dynamic [in providing a live management tool which supports improvement and looks to the future]; flexible [being readily applicable to different types of organisation and to units within those organisations]; innovative.
  (European Foundation for Quality Management 1999b)

In a study on self-assessment, Hillman (1994) has elaborated further on the benefits of the EFQM Model, noting:

- It is not a standard, but it allows for interpretation of all aspects of the business and all forms of organisation.
- Its widening use facilitates comparison between organisations. This provides the potential to learn from others in specific areas by using a common language.
- The inclusion of tangible results ensures that the focus remains on real improvement, rather than a preoccupation with the improvement process, i.e. it focuses on achievement not just activity.
- Training is readily available in the use of scoring for the model.
- It provides a repeatable basis that can be used for comparison over several years.
- The comprehensive nature and results focus, when broken down into discrete elements, helps develop a total improvement process specific for each organisation – it is a model for attaining a successful business.

Benefits derived from the implementation of the Excellence Model

The following provide the underpinning rationale for construction companies pursuing a competitive strategy through the application of the EFQM Excellence Model; the model is recognised as:

- providing a marketing focus;
- being a means of achieving a top-quality performance in all areas of the firm;
- providing valid operating procedures for all staff;
- allowing for the review of organisational self-assessment performance through providing a competitive weapon via a quality approach.

The model incorporates the RADAR concept based upon the Deming improvement philosophy depicted in *Figure 4.9*.

Constituent parts

- Plan: Identify customer needs and expectations, set strategic objectives.
- Do: Implement and operate processes.
- Think: Collect business results. Monitor and measure the processes, review and analyse.
- Act: Continually improve process performance.

Watson (2000) stated that 'the EFQM Model provided a truly service focused quality system which had an inbuilt mechanism for the attainment

of continued organisational improvement'. Wiele van der et al. (1997) identified that 'the criteria of the model helped managers to understand what
TQM means in relation to managing a company'.

The application of the model is simple, holistic, dynamic and flexible.
Further the model enhances senior management's understanding of TQM
(Watson and Seng 2001).

The generic model is depicted in *Figure 4.10* (Watson and Seng 2001) and
has been designed to be adopted or adapted in order to assist construction
organisations when engaged in the implementation of the EFQM EM.

EFQM EM *deployment as a project*

Watson (2002) described the main differences between management and
project management in that management usually consists of a set of tasks
that are repeated within a steady and reliable procedure. Project management, however, is related to the activities of a 'one off' specific project. A
project can be more efficiently and effectively managed if the Deming Control
Cycle is employed as depicted in *Figure 4.9*.

Harrison (1992) described the traditional form of management as not
being able to handle projects effectively. However, Watson (2002) 'emphasised that project management is a more challenging process', and he
described seven characteristics of project management. First, the 'role ends
when the project ends'. Second, 'start date, end date is difficult to predict'
due to the uniqueness of most projects. Third, the 'temporary [nature of]
project team'. Fourth, 'many different skills [are required in managing pro-
jects] ... in teams'. Fifth, 'costs [are] very difficult to estimate'. Sixth, 'work
often [new to team, again linked to the uniqueness of projects] ... has not
done before'. Seventh, 'time, cost and quality constraints'.

Harrison (1992) outlined the specific characteristics of project management
and why they are unique:

- projects are temporary [in nature and conducted] over a known duration;
- they involve several departments and companies;
- the complexity of integrating all activities, people and departments;
- the organisation structure of a project is unique and complex;
- different phases of the project [will] demand different [staff] groups for
  each phase, as a result relationships among staff [may become] unstable.

Furthermore, projects go through life-cycle stages (Conception, Definition,
Design, Execution, Commissioning). Therefore, quality, time and costs
have to be a prime consideration of the team and hence they will play a key
role in ensuring that each party involved in the project have well-defined
objectives.

Kerzner (2001) suggested that to be an effective project manager, and for
the deployment of EFQM EM [to be treated as a project], an individual must

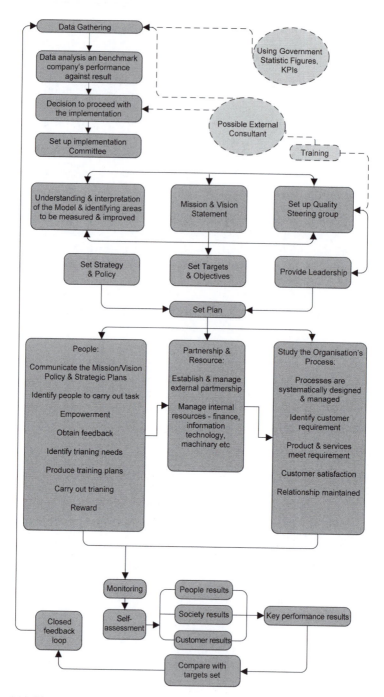

Figure 4.10 EFQM Excellence Model deployment

have management skills as well as technical skills. Kerzner (2001) proposed ten key skills:

- Team building skills: Managing cross functional teams and disciplines and being able to integrate them, into one group.
- Leadership skills: It is fundamental for a project manager to be able to manage and control the team. To do so a project manager must attain and demonstrate key personal characteristics such as 'innovate thinking', 'flexibility and change orientation' and 'good communication skills'.
- Conflict resolution skills: In projects conflict is inevitable; the result of unresolved conflict among groups in projects may lead to delays to the project itself. However, conflict can be encouraged in certain conditions, where it may lead to innovations. The role of the project manager is to be able to resolve conflict, by generating an environment where the objectives of the project are clear, and identifying the causes of conflicts.
- Technical expertise skills: It is important to be able to assess technical issues; however, a project manager should also use the skills, technical knowledge and competences of their team.
- Planning skills: A project manager has to demonstrate the ability to plan and control all activities.
- Organisational skills: Since the project manager's role is to integrate people from different departments, the project manager needs to understand how the organisation works and how to work with the organisation.
- Entrepreneurship skills: The project manager must consider the broad vision of the organisation. There are many more issues to consider, such as organisational growth.
- Administration skills: A project manager should be familiar with the basic skills of 'staffing', 'budgeting' and 'scheduling'.
- Management support: Usually a project manager is a linking pin between senior management and the rest of the project team members, who have to deliver the projects. Accordingly, a project manager should be capable of building a good working relationship with all parties; however, they are also entitled to the support of senior management, should the occasion arise where it is required.
- Resource allocation: A project manager needs to allocate the human and nonhuman resources in the most effective and efficient manner.

It has to be understood that the project manager has to manage up as well as down; this means they have to manage the client and senior staff of the organisation who may not be specifically involved in the project (deployment of EFQM EM).

Effective teams in deploying the EFQM EM

Dale (1999) proposed that the 'health' of a team is a significant factor that needs to be regularly assessed. However, he also stressed that it is not an easy task to evaluate the effectiveness of team working. Thus he suggested the following characteristics of an effective team:

- Everyone in the team is participating, making a contribution and is involved in appropriate actions, and through this they are achieving their personal potential in line with project goals.
- Relationships are open and team members trust and respect each other.
- Members listen closely to the views of other members of the team and have an open mind and maintain a positive attitude towards the rest of the team and the project.
- Everyone can express their views and ideas, with problems being addressed by the team, if appropriate.
- Members respect the operating procedures and principle of the team, and they own the team process.
- There is clarity and unity of focus on the project and all members know what is expected of them in achieving project goals.
- The TQM (EFQM EM) team leader has the ability to translate ideas into actions.

Watson (2000) noted the importance of teamwork as the core element for attaining cultural change within an organisation. In addition, Watson (2002) emphasised that TQM (EFQM EM) is based on involving all employees in achieving both the objectives of an organisation and their own personal objectives, such as improved salaries and enhanced promotional prospects. And most importantly to maximise the performance of their respective teams in the completion of project tasks.

Love et al. (2000) discussed how employees can be formed into 'Self-directed work teams', which were defined as a number of employees forming an effective operational group. They suggested developing 'in-house employee education sessions' which consist of specific classroom teamwork exercises formed of multi-disciplinary groups (managers, office staff, etc) in order to break down managerial and work sectional barriers. They also summarised the benefits of such workshop sessions as follows:

- Provides an improvement in communication links between work section members.
- Assists in breaking down work section barriers.
- Enables an improved awareness of how teamwork can help in achieving and exceeding corporate and personal targets.
- Highlights the significance and benefits of preparation and planning.
- Emphasises the need for effective time management.
- Underscores the importance of effective resource management.
- Helps the development of an internal/external customer culture.
- Can assist in the resolution of conflicts.

Ho (1999) addressed another important human factor within the TQM/EFQM EM implementation process; this is the establishment of 'quality control circles'. It consists of a small group of employees, which contribute to the improvement of the firm. They rely on using quality control methods to solve repeated occurring problems. Some of the methods utilised consist of cause-and-effect diagrams, histograms, scatter diagrams and quality control charts (Dale 1999).

Empowerment

Both individual and team empowerment are essential components in the successful application of the EFQM EM. Individual empowerment is vital for developing the self-esteem of the employee and to encourage individual motivation. While team motivation helps groups to deal more easily with difficult business environments, it also leads to the effective integration of group/team members (Dainty et al. 2002).

Dainty et al. also suggest that empowerment is comprised of three core elements:

- Clarity of vision and mission;
- Consistency of organisational goals and the alignment of systems directed towards these goals;
- An ongoing evaluation of the professional needs of employees; and congruence between corporate, management and employee goals.

Several benefits can be attained as a result of attaining effective empowerment in individuals and teams:

- Improved productivity and quality;
- Reduced operating costs;
- Allows for greater flexibility;
- Increases job satisfaction and hence the improved motivational aspects of all staff.

Teamwork through self-assessment

It has been established that the involvement of all employees is one of the fundamental components of TQM and hence a key component of the EFQM EM. The most appropriate method of involving all employees in a systematic review of their processes is through self-assessment application. The process of self-assessment will result in the identification of an organisation's strengths and potential improvement opportunities (Finn and Porter 1994).

Training

Dooley (1998) emphasised the importance of training as a way of improving the implementation of TQM/EFQM EM, noting that organisations must introduce and provide major improvements in TQM/EFQM EM training and educational programmes. Training needs to be well organised and clearly focused on the needs of the company and project demands. Dooley proposed that training should be practical and sometimes even experimental, and most importantly that the learned skills and competences must be implemented immediately in order for transfer of learning to occur. Further training programmes should develop specific objectives for learning, so assessment could be conducted by measuring how successful the training programme is against the determined objectives.

Another method that can assist in the effective application of TQM/EFQM EM is based on the establishment of a 'broad networking with other firms'. This can be accomplished in various ways such as conferences and monthly meetings, and professional body activities, such as continuing professional development (CPD) events. An obvious way of gaining new skills quickly for a construction company is to seek out and employ external consultants.

Training programmes must be consistent in content if offered throughout an organisation and should take into consideration any possible cultural differences that may exist between departmental boundaries (Reavill 1999).

Scoring the European Foundation for Quality Management Excellence Model

The following example is based upon a fictitious organisation and is only applied to the 'Leadership' enabler and 'Performance' results criteria (to engage in the whole process would require approximately 60 pages). Readers are therefore advised to obtain a copy of the full scoring document from EFQM, Brussels, or visit the website address noted on the reference pages at the end of this chapter.

Leadership defined by EFQM

How leaders develop and facilitate the achievement of the mission, vision, develop values required for long-term success and implement these via appropriate actions and behaviours, and are personally involved in ensuring that the organisation's management system is developed and implemented.

The above has been used to evaluate the fictitious organisation and a score allocated by (trained) senior management. The sub-areas are graded and a final score awarded, for example 'Approach' in this case has been allocated a score of 55 per cent, as depicted in *Table 4.4*.

In the example, Leadership is awarded a score of 55 per cent for Approach, 45 per cent for Deployment, 40 per cent for Assessment and Review and an overall score of 45 per cent. It is the overall score of 45 per cent that is carried forward to the scoring summary sheet (*Table 4.6*). However, before we engage in this process let us apply the scoring system to performance results (Table 4.5).

Table 4.4 Scoring the EFQM leadership

Elements	Scores Attributes	0%	25%	50%	75%	100%
Approach	**Sound:** – approach has a clear rationale – there are well defined and developed processes – approach focuses on stakeholder needs	No evidence or anecdotal	Same evidence	Evidence	Clear evidence	Comprehensive evidence
	Integrated: – approach supports policy and strategy – approach is linked to other approaches as appropriate	No evidence or anecdotal	Same evidence	Evidence	Clear evidence	Comprehensive evidence
		0% 5% 10% 15% 20% 25% 30% 35% 40% 45% 50% **55%** 60% 65% 70% 75% 80% 85% 90% 95% 100% (X at 55%)				
Deployment	**Implemented:** – approach is implemented	No evidence or anecdotal	Implemented in about 1/4 of relevant areas	Evidence	Implemented in about 3/4 of relevant areas	Implemented in all relevant areas
	Systematic: – approach is deployed in a structured way	No evidence or anecdotal	Same evidence	Evidence	Clear evidence	Comprehensive evidence
		0% 5% 10% 15% 20% 25% 30% 35% 40% **45%** 50% 55% 60% 65% 70% 75% 80% 85% 90% 95% 100% (X at 45%)				
Assessment and Review	**Measurement:** – regular measurement of the effectiveness of the approach, deployment is carried out	No evidence or anecdotal	Same evidence	Evidence	Clear evidence	Comprehensive evidence
	Learning: – learning activities are used to identify and share best practice and improvement opportunities	No evidence or anecdotal	Same evidence	Evidence	Clear evidence	Comprehensive evidence
	Improvement: – output from measurement and learning is analysed and used to identify, prioritise, plan and implement improvements	No evidence or anecdotal	Same evidence	Evidence	Clear evidence	Comprehensive evidence
		0% 5% 10% 15% 20% 25% 30% 35% **40%** 45% 50% 55% 60% 65% 70% 75% 80% 85% 90% 95% 100% (X at 40%)				
Overall Total		0% 5% 10% 15% 20% 25% 30% 35% 40% **45%** 50% 55% 60% 65% 70% 75% 80% 85% 90% 95% 100% (X at 45%)				

Performance

Key performance results

What the organisation is achieving in relation to its planned performance.

Key performance indicators

These measures are the operational ones used in order to monitor, understand, predict and improve the organisation's likely key performance outcomes, depending on the purpose and objectives of the organisation and its processes, as shown in *Table 4.5*.

It should be noted that when scoring against the criteria, management will find it most useful to consider areas of:

- Strengths: areas of good/best practice that could be disseminated throughout the organisation.
- Areas for improvement so that corrective actions can be employed.
- Evidence which supports the awarded percentage points.

Upon completion of the scoring related to the five enablers and four results (with sub-criteria) the scores are carried forward to the scoring summary sheet, *Table 4.6*.

In the example used, the scores for Leadership and Performance, along with a completed analysis, have been inserted so a final score can be obtained. This is shown in *Table 4.6*.

- Enter the score awarded to each criterion (of both sections 1 and 2 above).
- Multiply each score by the appropriate factor to give points awarded.
- Add points awarded to each criterion to give total points awarded for applications.

In order to put the score in the context of best practice it should be noted that the EFQM will conduct a site visit on an organisation obtaining over 500 points. Also the EFQM award for excellence is usually awarded to organisations obtaining a score between 750 and 850 points.

The scoring summary sheet provides a useful overall picture of the construction organisation. However, this book's authors have developed the data further to provide more detailed information for the host company. It would be very useful for a company to know the profile related to: Approach; Deployment; Assessment and review; Criteria; Results. This would allow the organisation to focus its efforts for improvement.

An example of the above approach follows. Note, average scores have been calculated for the noted areas and 'Results' have been divided into 'Results' and 'Scope', thus providing more detail, enabling more effective corrective actions to be deployed.

Table 4.5 Scoring the EFQM performance results

Elements	Scores Attributes	0%	25%	50%	75%	100%
Results	**Trends:** • trends are positive and/or there is sustained good performance	No results or anecdotal information	Positive trends and/or satisfactory performance on some results	Positive trends and/or sustained good performance on many results over the last 3 years	Strongly positive trends and/or sustained excellent performance on most results over at least 3 years	Strongly positive trends and/or sustained excellent performance in all areas over at least 5 years
	Targets: • targets are achieved • targets are appropriate	No results or anecdotal information	Favourable and appropriate in some areas	Favourable and appropriate in many areas	Favourable and appropriate in most areas	Excellent and appropriate in most areas
	Comparisons: • comparisons with external organisations takes place and results compare well with industry averages or acknowledged 'best in class'	No results or anecdotal information	Comparisons in some areas	Favourable in some areas	Favourable in many areas	Excellent in most areas and 'Best in Class' in many areas
	Causes: • results are caused by approach	No results or anecdotal information	Some results	Many results	Most results	All results. Leading position will be maintained
	TOTAL	0% 5%	10% 15% 20% 25% 30%	35% 40% 45% 50% 55% **X**	60% 65% 70% 75% 80%	85% 90% 95% 100%
Results	**Scope:** results address relevant areas	No results or anecdotal information	Some areas addressed	Many areas addressed	Most areas addressed	All areas addressed
	TOTAL	0% 5%	10% 15% 20% 25% 30%	35% 40% 45% 50% 55%	60% **65% X** 70% 75% 80%	85% 90% 95% 100%

(Again these figures have been used to evaluate the host organisation in order to allocate a score.)

OVERALL TOTAL		0% 5%	10% 15% 20% 25% 30%	35% 40% 45% 50% 55%	**60% X** 65% 70% 75% 80%	85% 90% 95% 100%

Table 4.6 Scoring summary sheet

Criterion Number	1	%	2	%	3	%	4	%	5	%
Sub-criterion	1a	**45**	2a	50	3a	60	4a	50	5a	45
Sub-criterion	1b	40	2b	50	3b	35	4b	50	5b	60
Sub-criterion	1c	45	2c	40	3c	40	4c	55	5c	60
Sub-criterion	1d	50	2d	30	3d	40	4d	40	5d	50
Sub-criterion			2e	45	3e	50	4e	35	5e	50
Sum		180		215		225		230		265
		÷4		÷5		÷5		÷5		÷5
Score awarded		45		43.2		45.2		46		53

Note: The score awarded is the arithmetic average of the percentage scores for the sub-criteria. If applicants present convincing reasons why one or more parts are not relevant to their organisation it is valid to calculate the average on the number of criteria addressed. To avoid confusion (with a zero score) parts of the criteria accepted as not relevant should be entered 'NR' in the table above.

2. Results criteria

Criterion Number	6		%	7		%	8		%	9		%
Sub-criterion	6a 50	× 0.75= 37.5		7a 60	× 0.75= 45		8a 50	× 0.25= 12.5		9a 65	× 0.50= 32.5	
Sub-criterion	6b 50	× 0.25= 12.5		7b 50	× 0.25= 12.5		8b **60**	× 0.75= 45		9b 55	× 0.50= 27.5	
Score awarded		50			57.5			57.5			60	

3. Calculation of total points

Criterion	Score awarded	Factors from model*	Points awarded
1. Leadership	45	× 1.0	45
2. Policy and Strategy	43.2	× 0.8	34.6
3. People	45.2	× 0.9	40.7
4. Partnerships and Resources	46	× 0.9	41.4
5. Processes	53	× 1.4	74.2
6. Customer Results	50	× 2.0	100.0
7. People Results	57.5	× 0.9	51.8
8. Society Results	57.5	× 0.6	34.5
9. Key Performance Results	60	× 1.5	90.0
Total Points Awarded			512.2

* Note these are the factors from the model

For ease of presentation, this data can be represented on a RADAR pentagonal profile developed by this book's authors. See *Figure 4.10*. Again this is presented in the example.

The pictorial representation of the RADAR pentagonal profile enables instantaneous understanding of the current state of the company. It is also a very quick and accurate method of benchmarking the host organisation.

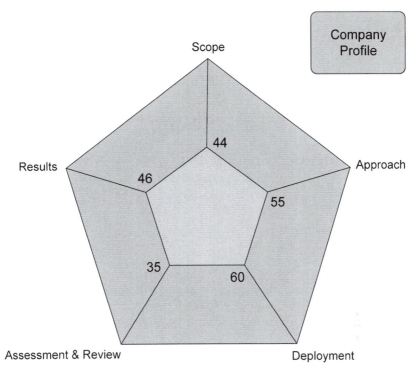

Figure 4.11 RADAR pentagonal profile
© Watson

Senior managers must remember that the self-evaluation process is designed to develop continuous improvement. Therefore, the benchmarking activity must be conducted on a regular basis so that corrective actions can be evaluated as part of a learning organisational activity.

Readers are asked to see EFQM's RADAR model, incorporated in their *Assessment Scoring Handbook*, for further details and for explanation of the scoring criteria.

The above scoring of the EFQM EM is an example of 'self-assessment benchmarking'.

Question 2 for the reader

The advantages of utilising EFQM EM's self-assessment methodology have been noted by Castka et al. (2003). Identify the advantages of EFQM EM's self-assessment methodology. An answer is provided at the end of the book.

Summary

The EFQM Excellence Model provides a valuable framework for addressing the key operational activities for construction-related organisations. It is useful because it enables a link to be made between people, organisational objectives and improvement processes, all encompassed under the umbrella of a continuous improvement philosophy.

The scoring methodology is simple to apply, but senior managers are advised to obtain some formal training, for themselves and their staff, before applying the model.

The model, when implemented, does provide detailed information for employing constant and consistent benchmarking/self-assessment activities, again focused on learning and improving.

Only a limited amount of detail can be provided within a text such as this; therefore, readers are requested to make specific reference to the EFQM *Assessment Scoring Handbook*. This document provides further information explaining details about the sub-assessment criteria such as 'Trends'.

Conclusions

Many construction organisations suffer from poor performance because of a combination of traditional organisational structures and management practices, while still operating within a dynamic competitive environment. Thus Love et al. (2000) advocated that TQM/EFQM EM could be the solution for such construction organisations by 'implementing the philosophical elements of total quality management'.

The EFQM Excellence Model is used as a tool for assisting in defining TQM in a way which construction senior management and employees can easily comprehend. However, to use any self-assessment method effectively, various elements and practices have to be in place and management needs to have had some experience of TQM to understand the questions underpinning the model on which self-assessment is being based (Wiele van der et al. 1997).

To be able to utilise the EFQM Excellence Model and conduct self-assessment optimally, there must be a trend for 'serious investment in resources' (Sommerville and Robertson 2000). And the most essential resource is people, since they are the real asset of any construction organisation. Wright (1997) pointed out that 'Indeed, it [EFQM EM] recognises that satisfied customers and staff are a far more powerful indicator of sustainable future success than financial measures alone'. In addition, Dale (1999) suggests that the performance of the organisation increases or decreases according to the way employees are treated and deployed. A 'Total quality approach is built on the commitment and motivation of people, achieved through relevant training, good communication and genuine consultation – all features of any effective [construction] organisation'.

Moreover, when it comes to deploying EFQM EM in public-sector organisations, many challenges should be expected. This is based on the findings of Osseo-Asare Jr. and Longbottom (2002) who suggested an important point to be taken into consideration is that 'private sector organisations have been more successful with the EFQM model than the public sector organisations'.

Within this text the European Foundation for Quality Management Excellence Model has been explained and its constituent parts established; further the advantages of its deployment are promoted. Construction organisations have much to gain by applying the EFQM EM.

Question 3 for the reader

The EFQM EM is based on, and supported by, specific concepts which are referred to as 'The Fundamental Concepts of Excellence'. Identify the Fundamental Concepts of Excellence. The answer is provided at the end of the book.

Case Study for the reader

Deploying EFQM EM at XYZ Construction Company

A new managing director has just been appointed to XYZ; the appointment has been made on the understanding that he will oversee the deployment of the EFQM EM within the company. However, the managing director has only a limited knowledge of the model. Yet he has to convince all company personnel of the deployment rationale. Therefore he has decided to engage external consultants to assist him. You have been appointed as external consultant and asked to prepare a presentation for the board of directors (based upon this chapter). Your presentation should consist of key bullet points. The bullet points should relate to the advantages of deploying the EFQM EM; however, you should also note any possible problematic issues of implementation.

When complete you can compare your list with that provided at the end of the book.

Further reading

British Quality Foundation, (1998). *Guide to the Business Excellence Mode – Defining World Class*. British Quality Foundation, London.

Crosby, P. B. (1979). *Quality if Free; The art of Making Quality Certain*, New York: McGraw-Hill.

Hermel, P. and Ramis-Pujol, J. (2003). An Evolution of Excellence: Some Main Trends. *TQM Magazine*, Vol. 15, No.4, pp. 230–43.

Powell, T. C. (1995). Total Quality Management as competitive advantage: a review and empirical study, *Strategic Management Journal*, Vol.16 pp. 15–37.

Watson, P. (2002). Developing an Efficient and Effective Control System. *Journal of the Association of Building Engineers*. Vol. 77 (3), February.

Wilkinson, A. and Witcher, B. (1991). Fitness for Use? Barriers to full TQM in the UK, *Management Decisions*, 29 (8), pp. 46–51.

References

Agere, S. and Jorm, N. (2000). *Designing Performance Appraisals*, 1st ed., Commonwealth Secretariat.

Ball, A., Bowerman, M. and Hawksworth, S. (2000). Benchmarking in local government under a central government agenda. *Benchmarking: An International Journal*, 7(1), pp. 20–34.

British Quality Foundation (1998). Self Assessment Techniques for Business Excellence. *Identifying Business Opportunities*, London.

Capon, N., Kay, M. and Wood, M. (1995). Measuring the success of a TQM programme. *International Journal of Quality & Reliability Management*, 12 (8), pp. 8–22.

Castka, P., Bamber, C. and Sharp, J. (2003). Measuring teamwork culture: the use of a modified EFQM model. *Journal of Management Development*, 22 (2), pp. 149–70.

Cherkasky, S. M. (1992). Total Quality for a Sustainable Competitive Advantage. *Total Quality Management Journal*, August, pp. 4–7.

Coleman, S. and Douglas, A. (2001). Where next for ISO 9000 companies? *Total Quality Management Journal*, 15 (2), pp. 88–92.

Conti, T. (1997). Optimizing Self-assessment, *Total Quality Management*, Vol. 9, Nos 2 & 3, S5–S15.

Curry, A. (1999). Innovation in public service management. *Managing Service Quality*, 9 (3), pp. 180–90.

Dainty, A., Bryman, A. and Price, A. (2002). Empowerment within the UK construction sector. *Leadership and Organization Development Journal*, 23 (6), pp. 333–42.

Dale, B. (1999). *Managing Quality*, 3rd ed. Blackwell Publishing.

Dooley, K. (1998). Perceptions of success and failure in TQM initiatives. *Journal of Quality Management*, 3 (2).

European Foundation for Quality Management (1999a). *Radar and the EFQM Excellence Model*, EFQM Press Releases & Announcements [online], last accessed on 12 June 2000 at: http://www.efqm.org/seflas.htm.

——(1999b). *Eight Essentials of Excellence – the fundamental concepts and their benefits.* Brussels Representative Office, Belgium.

——(2000a). *History of the EFQM*, [accessed 12 June 2000] at: http://www.efqm.org/history/htm.

——(2000b). *EFQM and Self-Assessment*, [accessed 12 June 2000] at: http://www.efqm.org/seflas.htm.

Finn, M. and Porter, L. (1994). TQM Self-assessment in the UK. *The TQM Magazine*. 6 (4), pp. 56–61.

Ham, J. (2005). The Creation of Meanings within Occupational Groups, *Proceedings of the International Commercial Management Symposium*, 7 April, pp. 52–62, The University of Manchester, UK.

Harrison, F. L. (1992). *Advanced Project Management, a Structured Approach* 3rd ed., Gower Publishing Company Ltd.

Harrison, M. (1993). *Operations Management*, London: Pitman Publishing.

Hendricks, K. B. and Singhal, V. R. (2001). Firm Characteristics, Total Quality Management, and Financial Performance, *Journal of Operations Management* 19, pp. 269–85.

Hickman, C. R. and Silva, M. (1989). *Creating Excellence*, London: Unwin Hyman Ltd.

Hillman, G. P. (1994). Making Self-assessment Success, *Total Quality Management*, Vol. 6 (3), pp. 29–31.

Ho, S. (1999). From TQM to business excellence. *Production Planning and Control*, 10 (1), pp. 87–96.

Johnson, G. and Scholes, K. (1999). *Exploring Corporate Strategy*. 5th ed., London: Prentice Hall.

Kanter, R. M. (1989). *When Giants Learn to Dance*, Reading: Cox Andwymann.

Kerzner, H. (2001). *Project Management: A systems approach to planning, scheduling, and controlling*, 7th ed., John Wiley & Sons, Inc.

Love, P., Li, H., Irani, Z. and Faniran, O. (2000). Total Quality Management and the Learning Organisation: a dialogue for change in construction. *Construction Management and Economics*, 18, pp. 321–31.

McAdam, R., Reid, R. and Saulters, R. (2002). Sustaining quality in the UK public sector quality measurement frameworks. *International Journal of Quality & Reliability Management*, 19 (5), pp. 581–95.

McKenna, E. (2000). *Business Psychology and Organisational Behaviour*. 3rd ed., Psychology Press Ltd.

Mohrman, S. A., Tenkasi, R. V., Lawler III, E. E. and Ledord Jr, G. E. (1995). Total Quality Management: practice and outcomes in the largest US firms, *Employee Relations*, 17(3), pp. 26–41.

Mundie, P. and Cottam, A. (1993). *The Management and Marketing of Services*. Kilbridge: Butterworth-Heinemann.

Mundy, K. (1992). Making the Right Choice. *Logistics Today*, September–October 1992.

Nunney, D. (1992). *Integrated Manufacturing*. The Department of Trade and Industry UK.

Oakland, J., Tanner, S. and Gadd, K. (2002). Best practice in business excellence. *Total Quality Management*, 13(8), pp. 1125–39.

Oakland, S. J. (1993). *Total Quality Management*, London: Butterworth Heineman.

——(1997). Interdependence and Cooperation: The Essentials of Total Quality Management, *Proceedings of the 2nd World Congress for Total Quality Management: The Quality Journal Sheffield*, UK. Vol. 8, Numbers 2 and 3, June.

Orsini, J. (2000). Troubleshooting your activities for excellence. *Total Quality Management*, 11(2), pp. 207–10.

Osseo-Asare Jr., A. and Longbottom, D. (2002). The need for education and training in the use of the EFQM model for quality management in UK higher education institutions. *Quality Assurance in Education*, 10(1), pp. 26–36.

Pitt, D. (1999). *Improving Performance Through Self-Assessment*. Business Development Manager, Wakefield and Pontefract Community Health. NHS Trust, Wakefield, UK.

Preston, J. (1993). *International Business*, London: Pitman.

Reavill, L. (1999). What is the future direction of TQM development? *The TQM magazine*, 11(5), pp. 291–98.

Ross, D. (1991). Aligning the Organisation for World Class Manufacturing. *Production and Inventory Management* Second Quarter 1991.

Spencer, B. A. (1994). Models of Organisational and Total Quality Management: A Comparison and Critical Evaluation. *Academy of Management Review* Vol. 19. No. 3. pp. 446–71.

Watson, M. (2002). *Managing Smaller Projects*. 2nd edition, Project Manager Today Publications.

Watson, P. (2000). Applying the European Foundation for Quality Management (EFQM) Model, *Journal of the Association of Building Engineers*, Vol. 75 (4), pp. 18–20.

Watson, P. and Chileshe, N., (2001). The Relationship between Organisational Performance and Total Quality Management within Construction SMEs. *Proceedings of CIB World Building Congress*, April 2–6, Wellington, New Zealand, (3), pp. 233–44.

Watson, P. and Seng, L. T. (2001). Implementing the European Foundation for Quality Management Excellence Model in Construction, *Construction Information Quarterly* 3 (2), Paper No 130.

Westlund, A. (2001). Measuring environmental impact on society in the EFQM system. *Total Quality Management*, 12 (1), pp. 125–35.

Wiele van der, A., Dale, B. D. and Williams, A. R. T. (1997). 'ISO 9000 Series Registration to Total Quality Management: the Transformation Journey', *International Journal of Quality Science*, June, Vol. 2 (4), pp. 236–52.

Wright, A. (1997). Public service quality: lessons not learned. *Total Quality Management*, 8 (5), pp. 313–22.

Yip, G. S. (1992). *Total Global Strategy*, New Jersey: Prentice Hall.

5 Quality and environmental management systems

Introduction

Industries, organisations, communities and consumers impact upon the environment in various ways and all have a shared responsibility for ensuring the integrity of the earth's environment for current and future generations. With regard to construction projects, environmental concerns often focus upon aspects such as air emissions (e.g. dust), waste disposal, noise pollution and water discharge.

The ISO 14001:2004 standard is an international standard for developing, supporting and auditing environmental management systems within all types of organisations. This chapter introduces environmental management systems and outlines the requirements placed upon an organisation by ISO 14001:2004.

Learning outcomes

Upon completion of this chapter the reader will be able to demonstrate:

- An awareness of the structure of ISO 14001 and the implementation process associated with the standard;
- An understanding of the conformance requirements and advocated benefits of the ISO 14001 standard;
- An understanding of the requirements, content and purpose of site waste management plans.

ISO 14001:2004 environmental management systems – requirements

The desire and need to implement environmental management systems has grown as organisations in various industries, sectors and geographic locations have come to recognise their impacts upon the environment. Many organisations have increasingly adopted socially responsible approaches to their business activity. ISO 14001:2004 is an international standard that enables organisations to be increasingly socially responsible by enabling environmental aspects of the business to be managed in a systematic, consistent, coherent, auditable and continually improving manner.

Like many management systems, the ISO 14001 standard is built upon Deming's 'Plan, Do, Check, Act' model. It expands Deming's model from four to five phases, these being Environmental Policy, Planning, Implementation and Operation, Checking and Corrective Action, and Management Review.

The International Organisation for Standardisation began consultation regarding an environmental standard in 1991 and in 1996 published the ISO 14001 standard which was revised in 2004. The current ISO 14000 series, of which 14001 is one standard, is outlined in *Table 5.1*.

This 14000 series of environmental management standards provides, amongst other things, a framework for developing, supporting and auditing environmental management systems within organisations. *Table 5.2* illustrates the standards that relate to the implementation of an environmental management system and locates them according to their optimal place in the PDCA cycle (ISO 2009).

ISO 14001 is the keystone standard of this series as it serves to specify a framework for an Environmental Management System against which an organisation can be certified by a third party. The standard is made up of 18 elements including the general requirements. The elements of the standard are:

- Requirement: 4.1 General Requirements
- Requirement: 4.2 Environmental Policy
- Requirement: 4.3.1 Planning – Environmental Aspects
- Requirement: 4.3.2 Legal and Other Requirements
- Requirement: 4.3.3 Objectives, Targets, and Programs
- Requirement: 4.4.1 Structure and Responsibility
- Requirement: 4.4.2 Competence, Training, and Awareness
- Requirement: 4.4.3 Communications
- Requirement: 4.4.4 EMS Documentation
- Requirement: 4.4.5 Control of Documents
- Requirement: 4.4.6 Operational Control
- Requirement: 4.4.7 Emergency Preparedness and Response
- Requirement: 4.5.1 Monitoring and Measurement
- Requirement: 4.5.2 Evaluation of Compliance
- Requirement: 4.5.3 Non-conformances, Corrective and Preventive Action
- Requirement: 4.5.4 Control of Records
- Requirement: 4.5.5 Internal Audit
- Requirement: 4.6 Management Review

Whilst the ISO 14001 standard specifies numerous requirements it does not specify absolute requirements for environmental performance. It requires organisations to commit to continual improvement and compliance with applicable legislation and regulations. As such, many organisations, whilst

Table 5.1 The ISO 14000 series

	Standard	Title
Environmental Management Systems Requirements	ISO 14001:2004	Environmental management systems: Requirements with guidance for use
Additional Guidance on Environmental Management Systems	ISO 14004:2004	Environmental management systems – General guidelines on principles, systems and supporting techniques
Site Assessments	ISO 14015:2001	Environmental management – Environmental assessment of sites and organizations
Environmental Labelling	ISO 14020:2000	Environmental labels and declarations – General Principles (Additional labelling standards are ISO 14024:1999, ISO/TR 14025:2000)
Environmental Performance Evaluation	ISO 14031:1999	Environmental management – Environmental performance evaluation – Guidelines (Additional performance evaluation standard is ISO 14032:1999)
Life Cycle Assessment	ISO 14040:2006	Environmental management – Life cycle assessment – Principles and framework (Additional life cycle assessment standards are ISO 14041, 14042, 14043, 14044, 14047, 14048, 14049)
Terms and Definitions	ISO 14050:2009	Environmental management – Vocabulary
Product Design	ISO 14062:2002	Environmental management – Integrating environmental aspects into product design and development
Environmental Communication	ISO 14063:2006	Environmental management – Environmental Communication – Guidelines and examples
Greenhouse Gas	ISO 14064-1:2006 ISO 14064-2:2006 ISO 14064-3:2006	Standards for quantification and reporting of greenhouse gas emissions and the validation and verification of greenhouse gas assertions
Audit Protocol (common for 9000 and 14000 series)	ISO 19011	

Table 5.2 The ISO family of standards and their Plan, Do, Check, Act relationship for the implementation of environmental management systems

Plan	Do	Check	Act
Environmental management system implementation	*Conduct life cycle assessment and manage environmental aspects*	*Conduct audits and evaluate environmental performance*	*Communicate and use environmental declarations and claims*
ISO 14050:2009 Environmental management – Vocabulary	ISO 14040:2006 Environmental management – Life cycle assessment – Principles and framework	ISO 14015:2001 Environmental management – Environmental assessment of sites and organizations (EASO)	ISO 14020:2000 Environmental labels and declarations – General principles
ISO 14001:2004 Environmental management systems – Requirements with guidance for use	ISO 14044:2006 Environmental management – Life cycle assessment – Requirements and guidelines	ISO 14031:1999 Environmental management – Environmental performance evaluation – Guidelines	ISO 14021:1999 Environmental labels and declarations – Self-declared environmental claims (Type II environmental labelling)
ISO 14004:2004 Environmental management systems – General guidelines on principles, systems and support techniques	ISO/TR 14047:2003 Environmental management – Life cycle impact assessment-Examples of application of ISO 14042	ISO 19011:2002 Guidelines for quality and/or environmental management systems auditing	ISO 14024:1999 Environmental labels and declarations – Type I environmental labelling – Principles and procedures
ISO/DIS 14005 Environmental management systems – Guidelines for the phased implementation of an environmental management system, including the use of environmental performance evaluation	ISO/TS 14048:2002 Environmental management – Life cycle assessment – Data documentation format		ISO 14025:2006 Environmental labels and declarations – Type III environmental declarations – Principles and procedures
			ISO/AWI 14033 Environmental management – Quantitative environmental information – Guidelines and examples

Figure 5.1 The ISO 14001 process

complying with ISO14001, have very different environmental management systems and performance.

Implementation of ISO 14001 requires ongoing management commitment if a system is to be developed, operated, reviewed, monitored and audited with a view to continuous improvement. The process of implementing ISO14001 is outlined in *Figure 5.1*.

Conformance with the ISO 14001:2004 process

In seeking to certify conformance with ISO 14001:2004 a thorough understanding of requirements is essential. For further guidance the reader is directed to Annex A of the 14001:2004 standard. Conformance to the ISO standard is certified via third party audit. Registration is conducted in two stages, with an informal preliminary or preparatory review.

This should consider all legislation relevant to the organisation's worksites. It should also identify all relevant environmental aspects that could arise from the organisation's worksites and should involve comprehensive consideration of all input processes and outputs.

An environmental policy must be put into place and this must be fully supported by top management who must ensure the policy:

a. is appropriate to the nature, scale and environmental impacts of its activities, products or services;

b. includes a commitment to continual improvement and prevention of pollution;

c. includes a commitment to comply with relevant environmental legislation and regulations, and with other requirements to which the organisation subscribes;

d. provides the framework for setting and reviewing environmental objectives and targets;

e. is documented, implemented and maintained and communicated to all employees;

f. is available to the public.

(ISO14001 Section 4.2)

The organisation needs to determine its primary environmental objectives, which are the focus of the improvement process and the company's environmental programme. This programme must specify goals or targets the means to achieve objectives. The scope of the environmental management system also needs to be identified and procedures and controls clearly established and expressed so as to enable the implementation of the policy and the achievement of targets. It is vital also that the policy, objectives and the environmental management system be communicated to the organisation's workforce.

The two formal stages of certification audit have defined requirements; these are outlined below.

Stage 1 registration audit – requirements

The stage one audit is conducted to enable the organisation to be ready for the stage two audit. The stage one audit entails:

- Conducting a documentation review to evaluate whether the organisation's documentation adequately covers all the requirements of the ISO standard.
- A review of the environmental aspects and impacts and their significance, and an evaluation of the organisation's site-specific conditions.
- A review of compliance with all applicable environmental legislation and regulations.
- Obtaining evidence that internal audits and management reviews are being planned and performed.
- A review of the organisation's nonconformance, preventive and corrective action system.
- Interviewing the organisation's management and workforce to assess their readiness for the stage two audit.
- An outcome of this stage one process is the identification of any nonconformities and opportunities for improvements. Nonconformities require a corrective-action response from the organisation and such aspects of concern require rectification and resolution prior to the carrying out of the stage two audit.

Stage 2 – registration audit

The stage two audit reviews the implementation of all the elements of the ISO 14001 standard by the organisation. Documentation and records to support the implementation are reviewed. This audit serves to assess the organisation's adherence to its own prescribed policies, objectives and procedures and to ascertain conformance to the requirements of the ISO 14001 standard. The auditor documents major and minor nonconformances and opportunities for improvement and makes a recommendation regarding certification registration. The organisation is expected to respond to minor and major nonconformance with a corrective-action plan.

Whilst certification audits are required every three years, the organisation's environmental management system requires ongoing comprehensive review to ensure that:

- it is effective in operation;
- there is progress of planned continual improvement activities;
- it is meeting specified objectives and goals;
- it is performing in accordance with relevant legislation regulations and standards.

Such ongoing surveillance audits facilitate management review and evaluation of the environmental management system to enable informed change and continuous improvement.

Table 5.3 presents a useful 'self-audit checklist' for the implementation of ISO 14001:2004. This checklist presents a series of questions, aligned with each section of the standard. This checklist is not exhaustive and should be used as one means of supporting the checking and management review of the system.

Advocated benefits of ISO 14001

A number of benefits can be associated with the implementation of an environmental management system such as ISO 14001. These benefits include:

- assisting an organisation in managing and reducing its impact upon the environment;
- enabling the organisation to meet its corporate social responsibility commitments;
- reducing costs incurred by the organisation due to reductions in, and minimisation of, waste and energy use;
- ensuring that the organisation meets both existing and future environmentally related regulations and requirements;
- increasing the opportunity for inclusion on tender shortlists, as accreditation can be a client prerequisite for such inclusion;

- enabling an organisation to obtain a competitive advantage;
- facilitating an increase in confidence in the organisation held by external stakeholders such as financiers and investors;
- enhancing the reputation of the organisation with consumers, clients and potential clients;
- providing a clear basis for the structured training of employees;
- reducing the potential for loss and expense arising from environmental claims due to the considered implementation of control measures and appropriate procedures;
- being a driver for continuous improvement within the organisation.

Table 5.3 A Self–audit for implementation of ISO14001:2004

4.1 General Requirements

Has the organisation established, documented, implemented, maintained and continually improved an environmental management system that is consistent with the requirements contained in ISO 14001 and has the organisation determined how it fulfils these requirements?

4.2 Environmental Policy

Does the organisation have an environmental policy that has been defined by top management?
Is the environmental policy appropriate for the nature / scale / environmental impacts of the organisation's activities / products / services?
Is a commitment to continual improvement and prevention of pollution included within the environmental policy?
Does the environmental policy include a commitment to comply with relevant legal requirements and with other requirements to which the organisation subscribes which relate to environmental aspects?
Is a framework for setting and reviewing environmental objectives and targets provided by the environmental policy?
Is the environmental policy documented, implemented, maintained?
Is the environmental policy communicated to all persons working for or on behalf of the organisation?
Is the environmental policy made available to the public?

4.3 Planning

4.3.1 Environmental Aspects

Has the organisation established, documented, implemented and maintained procedures to identify the environmental aspects of its activities/ products/ services within the defined scope of the environmental management system that it can control and furthermore, those that it can influence when taking into account planned or new developments, or new or modified activities/ products / services?
Has the organisation established, documented, implemented and maintained procedures to determine which of its aspects have or can have a significant impact on the environment?
Does the organisation ensure that significant environmental aspects are taken into account when establishing, documenting, implementing and maintaining its environmental management system?

4.3.2 Legal and other requirements

Has a procedure been put in place and maintained to identify and deliver access to legal requirements and other requirements subscribed to by the organisation that are relevant to the organisation's environmental aspects?

4.3.3 Objectives, Targets and Programmes

Has the organisation developed and maintained documented environmental objectives and targets, for all relevant functions and levels of the organisation?
Are the objectives and targets of the organisation consistent with the environmental policy, measurable and committed to the prevention of pollution, continuous improvement and compliance with legal and other subscribed requirements?
In establishing and reviewing its objectives, has the organisation considered:

- Legal and other requirements?
- Significant environmental aspects?
- Technological options?
- Financial, operational, and business requirements?
- The views of interested stakeholders?

Have environmental management programmes been established and maintained for achieving the objectives and targets of the organisation?
Do environmental management programmes designate responsibility for achieving objectives and targets at each relevant function and level of the organisation?
Do environmental management programmes define the means and timeframe by which they are to be achieved?

4.4 Implementation and Operation

4.4.1 Resources, Roles, Responsibility and Authority

Have essential resources (human, financial, infrastructure, technological) been provided by management to facilitate the implementation, control and improvement of the environmental management system?
Has the organisation defined, documented, and communicated the roles, responsibilities, and authorities to facilitate effective environmental management? How is this achieved?
Have the organisation's senior management appointed specific management representatives who have defined roles, responsibilities, and authority to:

- Ensure that the environmental management system (EMS) requirements are established, implemented and maintained in accordance with the 2004 International Standard.
- Report to senior management regarding performance of the EMS for the purpose of review and to facilitate improvement of the EMS.

4.4.2 Competence, Training and Awareness

How does the organisation ensure that those persons carrying out tasks that could cause significant environmental impacts are competent and possess appropriate education, training, and/or experience? Are records retained?
Has the organisation identified the training needs for all personnel whose work may create a significant impact upon the environment? How is this done?

Table 5.3 (continued)

Has the organisation established and maintained procedures to ensure that all its employees are aware of:

- The importance of conformance with the environmental policy and procedures and with the requirement of the environmental management system (EMS)?
- The significant environmental impacts, actual or potential, of work activities and the environmental benefits of improved personal performance?
- Roles and responsibilities in achieving conformance with the environmental policy, procedures and the requirements of the EMS?
- The potential consequences of departure from specified operating procedure?

4.4.3 Communication

With regard to environmental aspects and environmental management system, has the organisation established and maintained procedures to ensure:

- Internal communication between the various levels and functions of the organisation?
- Proper receiving, documenting, and responding to relevant communication from external interested parties?

Has the organisation considered and documented its decision regarding the processes for external communication regarding its significant environmental aspects?

4.4.4 Documentation

Has the organisation established and maintained the following documentation

- An environmental policy, including objectives and targets?
- An environmental management systems manual including:
- Description of the scope of the EMS?
- Description of the core elements of the management system and how these elements interaction?
- Documentation relating to records and control of processes?

4.4.5 Control of Documents

Has the organisation established and maintained procedures for controlling all documents required by this International Standard to ensure:

- Changes and the current revision status of documents are identified?
- The current versions of relevant documents are available at all points of use?
- Documents of external origin relative to the EMS are identified and controlled?
- Obsolete documents are promptly removed from all points of issue and points of use, or assured against unintended use?
- Obsolete documents are retained for legal with knowledge preservation purposes being suitably identified?
- Documents are legible, dated (with dates of revision), clearly identified, maintained in an orderly manner and retained for a specified period?
- Have procedures and responsibilities concerning the creation and modification of the various types of documents been clearly established and maintained?

4.4.6 Operational Control

Has the organisation identified and planned those operations and activities that are associated with the identified significant environmental aspects in line with its policy, objectives and targets?

Has the organisation planned these control activities, including maintenance, in order to ensure that they are carried out under specified conditions by:

- Establishing and maintaining documented procedures to cover situations where their absence could lead to deviations from the environmental policy and the objectives and targets?
- Stipulating operation criteria in the procedures?
- Establishing, implementing and maintaining procedures related to the identified significant environmental aspects of goods and services used by the organisation and communicating relevant procedures and requirements to suppliers and contractors?

4.4.7 Emergency Preparedness and Response

Are procedures established and maintained to identify potential for and to respond to accidents and emergency situations, and to prevent and mitigate the environmental impacts that may be associated with them?

Has the organisation reviewed and revised, where necessary, its emergency preparedness and response procedure, in particular, after the occurrence of incidents or emergency situations?

Where practicable, has the organisation periodically tested its emergency preparedness and response procedures?

4.5 Checking

4.5.1 Monitoring and Measuring

Has the organisation established and maintained procedures to regularly monitor and measure the key characteristics of its operations and activities that can have a significant impact on the environment?

Does monitoring and measurement include the recording of information to track performance, relevant operational controls and conformance with the organisation's environmental objectives and targets?

Is equipment calibrated by the organisation and are records maintained in accordance with the organisation's procedures?

4.5.2 Evaluation of Compliance

Has a procedure been established and maintained for periodically evaluating compliance with applicable environmental legislation and regulations? And are records maintained of these evaluations?

Has a procedure been established and maintained for periodically evaluating compliance with requirements to which the organisation subscribes? And are records maintained of these evaluations?

4.5.3 Nonconformity, Corrective Action and Preventative Action

Have procedures been established, implemented and maintained for handling and investigating nonconformances, determining root cause, taking action to mitigate any impacts caused, and for initiating and completing corrective and preventive action to prevent recurrence?

Has the organisation implemented and recorded any changes in the recorded procedures resulting from corrective and preventive action?

Table 5.3 (continued)

Has a process been implemented and maintained to review the effectiveness of corrective and preventive actions taken been implemented by the organisation? When corrective and preventive actions are taken to eliminate the causes of actual and potential nonconformances, are they appropriate to the magnitude or problems and commensurate with the environmental impact encountered?

4.5.4 Control Records

Are records maintained, as appropriate to the system and the organisation, so as to demonstrate conformance to the requirements of this International Standard?
Has the organisation established and maintained procedures for the identification, maintenance, and disposition of environmental records, including:

- Training records?
- The results of audits (EMS audits)?
- The results of reviews (Management Reviews)?

Are the organisation's records legible, identifiable, and traceable to the activity / product / service involved?
Are the environmental records stored and maintained in such a way that they are readily retrievable and protected against damage, deterioration, or loss?
Are the retention times for records established and recorded?

4.5.5 Internal Audit

Has the organisation established and maintained programme(s) and procedure(s) for the carrying out of periodic audits of the environmental management system in order to:

- Determine whether or not the environmental management system conforms to planned arrangements for environmental management including the requirements of this International Standard and has been properly implemented and maintained?
- Provide information on the results of audits to management?

Is the organisation's audit programme, including any schedule, based on the environmental importance of the activity concerned and the results of previous audits? Does the audit procedure cover the audit scope, frequency, and methodologies, as well as the responsibilities and requirements for conducting audits and reporting results?

4.6 Management Review

Does the organisation's top management, at planned intervals, review the environmental management system to ensure its continuing suitability, adequacy and effectiveness?
Are the management reviews documented and retained?
Does each management review address the possible need for changes to policy, objectives, and other elements of the environmental management system, in light of:

- Environment management system audit results?
- Changing circumstances?
- The commitment to continual improvement?

A case study of site waste management plans – environmental management requirements for English construction projects

The ISO 14001:2004 standard does not specify absolute environmental management requirements for organisations and projects but legislation and regulations do. Within England one such regulation concerns the management of waste on construction projects. In June 2008 HM Government in association with the Strategic Forum for Construction published a 'Strategy for Sustainable Construction'. This strategy set out an aspiration to lead the world in sustainable construction. It declared that environmental targets could not be met without 'dramatically reducing the environmental impacts of buildings and infrastructure construction; we have to change the way we design and build'.

The targets of the strategy presented a challenge for improving sustainability within the construction industry. They also underlined the agenda for very real improvements in environmental management within the English construction industry. The overarching targets and headings of the strategy are presented in the summary *Table 5.4*.

Table 5.4 Strategy for sustainable construction

	Headings	Overarching targets
The 'Means'	Procurement	To achieve improved whole life value through the promotion of best practice construction procurement and supply side integration, by encouraging the adoption of the Construction Commitments in both the public and private sectors and throughout the supply chain.
	Design	The overall objective of good design is to ensure that buildings, infrastructure, public spaces and places are buildable, fit for purpose, resource efficient, sustainable, resilient, adaptable and attractive. Good design is synonymous with sustainable construction. Our aim is to achieve greater use of design quality assessment tools relevant to buildings, infrastructure, public spaces and places.
	Innovation	To enhance the industry's capacity to innovate and increase the sustainability of both the construction process and its resultant assets.
	People	An increase in organisations committing to a planned approach to training (e.g. skills pledges; training plans; Investors in People or other business support tools; Continuous Professional Development (CPD); life-long learning). Reduce the incidence rate of fatal and major injury accidents by 10% year on year from 2000 levels.
	Better regulation	A 25% reduction in the administrative burdens affecting the private and third sectors, a 30% reduction in those affecting the public sector by 2010.

Table 5.4 (continued)

	Headings	Overarching targets
The 'Ends'	Climate Change Mitigation	Reducing total UK carbon dioxide (CO_2) emissions by at least 60% on 1990 levels by 2050 and by at least 26% by 2020. Within this, Government has already set out its policy that new homes will be zero carbon from 2016, and an ambition that new schools, public sector non-domestic buildings and other non-domestic buildings will be zero carbon from 2016, 2018 and 2019 respectively.
	Climate change adaptation	To develop a robust approach to adaptation to climate change, shared across government.
	Water	To assist with the Future Water vision to reduce per capita consumption of water in the home through cost-effective measures, to an average of 130 litres per person per day by 2030, or possibly even 120 litres per person per day depending on new technological developments and innovation.
	Biodiversity	That the conservation and enhancement of biodiversity within and around construction sites is considered throughout all stages of a development.
	Waste	By 2012, a 50% reduction of construction, demolition and excavation waste to landfill compared to 2008.
	Materials	That the materials used in construction have the least environmental and social impact as is feasible both socially and economically.

The requirement for site waste management plans

In February of that same year the Site Waste Management Plans Regulations 2008 were laid before parliament and came into force on 6 April 2008. These regulations place specific environmental management requirements on construction projects delivered in England with an estimated cost of greater than £300,000 (excluding VAT) that they must prepare a site waste management plan (SWMP). Such SWMPs must be prepared prior to commencement of the project. It is the client's responsibility to prepare the initial pre-construction SWMP and then pass this to the appointed principal contractor. The principal contractor is responsible for developing and updating the plan before returning it to the client at the end of the project.

Regulation 6 stipulates the requirements of a site waste management plan:

1. A site waste management plan must identify:

   a) the client;
   b) the principal contractor; and
   c) the person who drafted it.

2. It must describe the construction work proposed, including:

   a) the location of the site; and
   b) the estimated cost of the project.

3. It must record any decision taken before the site waste management plan was drafted on the nature of the project, its design, construction method or materials employed in order to minimise the quantity of waste produced on site.

4. It must:

   a) describe each waste type expected to be produced in the course of the project;
   b) estimate the quantity of each different waste type expected to be produced; and
   c) identify the waste management action proposed for each different waste type, including re-using, recycling, recovery and disposal.

5. It must contain a declaration that the client and the principal contractor will take all reasonable steps to ensure that:

   a) all waste from the site is dealt with in accordance with the waste duty of care in section 34 of the Environmental Protection Act 1990(3) and the Environmental Protection (Duty of Care) Regulations 1991(4); and
   b) materials will be handled efficiently and waste managed appropriately.

With regard to the updating of the SWMP, Regulation 7 stipulates requirements for a project of £500,000 or less:

1. If the project has an estimated cost of £500,000 or less, whenever waste is removed from the site the principal contractor must record on the site waste management plan:

   a) the identity of the person removing the waste;
   b) the types of waste removed; and
   c) the site that the waste is being taken to.

2. Within three months of the work being completed the principal contractor must add to the plan:

   a) confirmation that the plan has been monitored on a regular basis to ensure that work is progressing according to the plan and that the plan was updated in accordance with this regulation; and
   b) an explanation of any deviation from the plan.

3. Failure to comply with this regulation is an offence.

With regard to the updating of the SWMP, Regulation 8 stipulates requirements for a project worth more than £500,000:

1. If the project has an estimated cost greater than £500,000 the principal contractor must update the site waste management plan in accordance with this regulation.

2. When any waste is removed the principal contractor must record on the plan:

   a) the identity of the person removing the waste;
   b) the waste carrier registration number of the carrier;
   c) a copy of, or reference to, the written description of the waste required by section 34 of the Environmental Protection Act 1990; and
   d) the site that the waste is being taken to and whether the operator of that site holds a permit under the Environmental Permitting (England and Wales) Regulations 2007 or is registered under those Regulations as a waste operation exempt from the need for such a permit.

3. As often as necessary to ensure that the plan accurately reflects the progress of the project, and in any event not less than every six months, the principal contractor must:

   a) review the plan;
   b) record the types and quantities of waste produced;
   c) record the types and quantities of waste that have been:

      i) re-used (and whether this was on or off site);
      ii) recycled (and whether this was on or off site);
      iii) sent for another form of recovery (and whether this was on or off site);
      iv) sent to landfill; or
      v) otherwise disposed of; and

   d) update the plan to reflect the progress of the project.

4. Within three months of the work being completed the principal contractor must add to the plan:

   a) confirmation that the plan has been monitored on a regular basis to ensure that work is progressing according to the plan and that the plan was updated in accordance with this regulation;
   b) a comparison of the estimated quantities of each waste type against the actual quantities of each waste type;
   c) an explanation of any deviation from the plan; and
   d) an estimate of the cost savings that have been achieved by completing and implementing the plan.

5. Failure to comply with this regulation is an offence.

Best practice in site waste management plans

The Waste and Resources Action Programme (WRAP) provide worthwhile tools and guidance to support and facilitate good practice in the use of site waste management plans. WRAP, a not-for-profit company established in 2000 and backed by government funding from England, Scotland, Wales and Northern Ireland, provide free access to online SWMP tools and know-how regarding the use of recycled materials and the practice of reclaiming and recycling materials from construction projects.

With regard to SWMPs, WRAP outline practical actions to deliver good and best practice. This is presented in *Figure 5.2*.

Further to the flowchart, WRAP provide a free 'SWMP Template' that has been developed to assist in the use of SWMPs and enable projects to go beyond compliance and achieve good and best practice. The template addresses the construction project cycle as a series of six stages:

1. **policy and setup**: administration details and setting targets;
2. **preparation and concept design**: initial concept and design decisions to reduce waste;
3. **detailed design**: waste forecasts and waste reduction actions;
4. **pre-construction**: waste carriers, waste destinations, waste management and recovery actions;
5. **construction**: recording actual waste movements, and;
6. **post completion and use**: KPIs, reporting, comparing actual quantities with estimates and the declaration.

(WRAP 2009b)

WRAP's practical guidance for achieving compliance, good and best practice is presented in *Table 5.5*.

Positive outcomes of construction project site waste management

The implementation of site waste management plans on construction projects can deliver positive environmental and economic outcomes. *Table 5.6* illustrates site waste management strategies implemented by three different organisations on construction projects and the outcomes achieved.

Furthermore the Building Research Establishment (2006) has estimated that every ten tonnes of waste reduction saves around five tonnes of CO_2 equivalent. It is clear that an increase in materials resource efficiency, the use of recycled materials and a reduction in waste materials going to landfill has a positive environmental impact.

The delivery of good practice in site waste management also brings economic benefits. According to WRAP (2010) cost savings can be achieved through the implementation of effective site waste management.

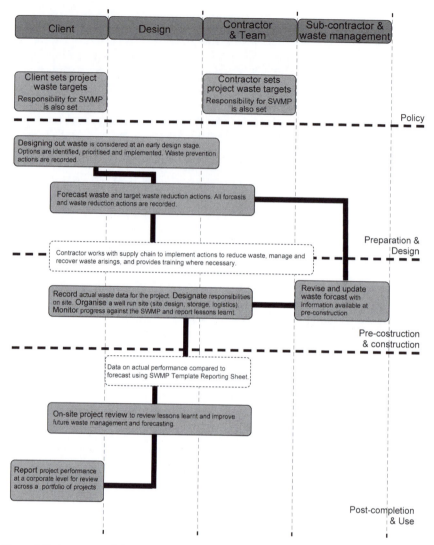

Figure 5.2 Actions to deliver good and best practice through a site waste management plan

Tables 5.6 and 5.7 identify cost savings achieved due to waste management good practice by various categories of project type and trade package.

Summary

This chapter has briefly outlined the structure of ISO 14001 and the implementation process associated with the standard. Conformance requirements and advocated benefits of the ISO 14001 standard have also been

Table 5.5 Six step practical guidance for achieving standard, good and best practice in construction project waste management

1.0 Policy

Step 1.1	Explanation	Practice level	How to achieve
Policy / target setting	At this early stage it is advisable that high-level targets are set which will govern and inform company strategy. These targets will then be incorporated into each construction project as they progress along the project lifecycle (and through the RIBA stages).	Standard	Set high level qualitative aspirational policy goals for company performance on reducing waste arisings and increasing waste recovery.
		Good	Insert quantified company-wide targets for reducing waste arisings and increasing waste recovery into company policy documents.
		Best	Process to insert quantified project specific waste reduction targets based on industry Best Practice benchmarks or previous project experience for reducing waste arisings and increasing waste recovery into company policy documents.

Step 1.2	Explanation	Practice level	How to achieve
Responsibilities (for the SWMP)	There are a number of required responsibilities for early stage coordination of the site waste management plan (SWMP).	Standard	Meet requirements for identifying the client, principal contractor and person drafting the site waste management plan.
		Good	Involve all members of the project team and ensure everyone knows about SWMP and how it affects them.
		Best	Include SWMP responsibilities as an agenda item at project team meetings, ensuring all team members are involved and contribute to project waste reduction and recovery actions.

2.0 Preparation and Concept Design

Step 2.1	Explanation	Practice level	How to achieve
Designing out waste	There are numerous opportunities to reduce waste during the design process. Designing out waste before it arises is one of the most efficient ways to reduce project waste arisings.However, as such decisions need to be taken early, engagement with the design team early on in the life of a project is key.	Standard	Capture decisions made that may have an impact on waste. These decisions may not have been taken with waste reduction in mind, but may have an effect on project waste arisings nonetheless.
		Good	Discuss with the project team at an early design stage how it might be best to reduce waste arisings through making changes to the design.
		Best	Systematically identify, prioritise and implement waste reduction actions at the design stage. Consider cost, programme and waste reduction potential.

Table 5.5 (continued)

3.0 Detailed Design

Step 3.1	Explanation	Practice level	How to achieve
Estimate waste arisings	Estimating waste arisings involves identifying and recording the amount and destination of each waste stream that will be generated on site. The earlier in the project life cycle that waste streams are estimated, the more opportunity there will be to prevent their creation.	Standard	Standard practice is to estimate waste arisings at the pre-construction stage.
		Good	Forecast waste arisings for each component using industry data.
		Best	Forecast waste arisings for each component using modified wastage rates based on past company experience.

Step 3.2	Explanation	Practice level	How to achieve
Target waste reductions	This step involves identifying and recording waste reduction methods to reduce the quantity of waste estimated in Step 3.1.	Standard	Identify waste management action for each of the different waste types forecast to arise on the construction project, including re-using, recycling, recovery and disposal.
		Good	Target waste arisings for each construction component using industry standard actions.
		Best	Target waste arisings for each construction component. As an example these actions could be to target accurate ordering (accurate material requirements, realistic wastage rates), logistics planning (delivery strategy, adequate storage, efficient movement of materials to the workface) or installation elements (efficient working and installation and storage of offcuts for reuse).

4.0 Pre-construction

Step 4.1	Explanation	Practice level	How to achieve
Forecast residual waste	In addition to designing out waste at (Step 2.1), and estimating outline waste arisings (Step 3.1), it is required to forecast residual waste arisings before going to site. This final residual waste forecast is the last and most detailed waste forecast that is done before site mobilisation. Once this final waste forecast is completed, waste management and recovery options can be implemented to ensure the waste is recycled, reused or recovered.	Standard	Forecast waste according to general estimates, fulfilling requirement to identify each waste type expected to be produced in the course of the project.
		Good	Good practice relates to forecasting waste arisings at the detailed design stage. Refer to Step 3.1. Good practice for Step 4.1 relates to forecasting residual waste arisings in conjunction with the principal contractor and agreeing the waste reduction and recovery standards to be achieved on the project.
		Best	Building on Good Practice, hold talks with the rest of the supply chain (waste management contractors, sub-contractors) to determine waste reduction and recovery actions for the project.

Step 4.2	Explanation	Practice level	How to achieve
Management of waste	This step relates to the efficient management of waste once it has been created on site. Step 4.2 which deals with the management of waste on site should be implemented in line with any targets identified in sections 1.0, 2.0 and 3.0 above. Offcuts should be stored safely on site for r e-use.	Standard	Identify waste management action for each waste stream.
		Good	Identify recycling and recovery options for each waste stream for which recycling and recovery is viable.
		Best	Maximise opportunities for resource efficiency through following the waste hierarchy (prevention, minimisation, reuse, recycling, recovery, disposal).

Table 5.5 (continued)

Step 4.3	Explanation	Practice level	How to achieve
Training	It is a requirement that all site workers are trained on the site waste management plan, providing information on how it affects them. Training prospects should be seen as opportunities to engage with the supply chain and gain buy-in from them – as it will be the supply chain who will be able to significantly contribute to any project=resource efficiency targets.	Standard	The principal contractor should provide training to every construction worker needed for the particular work to be carried out within the terms of the site waste management plan. This can be in the form of toolbox talks.
		Good	Building on standard practice, provide bespoke training to all subcontractors and identify waste reduction actions where they can contribute.
		Best	Building on good practice and share experience from previous projects or sites. Use the training exercise to inform continual improvement.

5.0 Construction

Step 5.1	Explanation	Practice level	How to achieve
Responsibilities (on site)	Once the SWMP has been developed it must be implemented on site. This Step outlines how to assign responsibility for ensuring the SWMP is delivered.	Standard	Meet requirements for identifying the client, principal contractor and person drafting the site waste management plan.
		Good	Waste champion is appointed for the whole site.
		Best	Building on Good Practice, individuals and subcontractors should be made responsible for specific waste streams, with the waste champion holding these project members to account.

Step 5.2	Explanation	Practice level	How to achieve
Site design, storage and logistics	Space permitting, key waste streams should be segregated. The segregation scheme should include appropriate training, monitoring and enforcement with clear signage and using the National Colour Coding Scheme.	Standard	Meet requirement that all waste from the site is dealt with in accordance with the Environmental Protection Act and Environmental Protection (Duty of Care) Regulations.
		Good	Before work starts on site consider layout and skip locations. Use segregated containers at the workface.
		Best	Ensure separate containers are provided for Hazardous Waste, material storage areas are clearly located and signed or arrange for just in time delivery and prevent double handling.

Step 5.3	Explanation	Practice level	How to achieve
Monitoring	Monitoring progress against the actions in the site waste management plan more often than every six months can inform ongoing site achievement of the planned waste reduction and recovery actions. It can be part of the live review process and inform continual improvement. Once data is collected, it will form a baseline against which clients can evaluate and improve on resource efficiency performance. Step 5.3 should therefore be linked with Step 6.2.	Standard	Monitor and update the site waste management plan not less than every six months.
		Good	Principal contractor to review the construction schedule and set appropriate project review and monitoring dates with the client.
		Best	Building on Good Practice, review site progress against the site waste management plan and implement changes to revise site activities based on performance where necessary.

Table 5.5 (continued)

Step 5.4	Explanation	Practice level	How to achieve
Reporting	Reporting is an integral part of the site waste management plan process. Good and best practice relate to recording and reporting waste arisings in increasing levels of detail. WRAP provide a method note that defines the standard by which the construction industry has agreed to record and report waste arisings. The link to this guidance is listed in the 'guidance' section opposite.	Standard	Ensure the site waste management plan is kept at the site, and that the plan is available for two years after completion of the construction project.
		Good	Report waste generation, recovery and disposal arising by construction phase (construction, demolition and excavation).
		Best	Report lessons learnt through the project, including the good and best practice levels achieved.

6.0 Post-completion

Step 6.1	Explanation	Practice level	How to achieve
On-site project review	The on-site project review is an opportunity for the site project team to review their progress post completion. Good and best practice items relate to the process of continuous review and learning.	Standard	Meet requirements to compare site waste management plan forecast versus actual performance, and record any deviations from the plan.
		Good	Building on Standard Practice, review the site waste management plan to identify any improvements that could have been made (e.g. to improve waste reduction or recovery, or the accuracy of the forecast).
		Best	Building on Good Practice, hold a post completion project team meeting to debrief and learn lessons from the site waste management plan process that can be used to inform future practice.

Step 6.2	Explanation	Practice level	How to achieve
Corporate level review	The corporate level review uses the SWMPs produced on individual sites to compare construction projects against company baseline performance. If a baseline does not exist, then the first project will become the baseline against which performance in future projects will be measured against.	Standard Good Best	Meet requirements to compare site waste management plan forecast versus actual performance, and record any deviations from the plan. Record project performance in the following areas: cost savings achieved, total waste arisings, total waste to landfill, total waste reductions achieved and recycled content used. Use data collected in Step 6.1 standard practice to benchmark performance across your portfolio of projects, using the data to inform continual improvement. Using the data gathered and lessons learnt, set company policy on expected metrics (cost savings, waste arisings, waste reductions, total waste to landfill) for similar project types going forward. Integrate lessons learnt into corporate construction procedures.

(extracted from WRAP 2009 (c))

Table 5.6 Outcomes of the implementation of construction site waste management

Project	Waste Management Strategy	Outcomes
A £4 million, 16 week fast-track programme – refurbishment of a clothing store.	• Materials segregated on site and placed into hippo bags, collected 10 at a time – due to lack of site space for storage and waste.	95% of waste was recycled (600 tonnes).
Design and build of a 90-bed hotel and office complex.	• Site employees trained to be aware of best practice. • Segregation of site waste – general, inert, metal, timber, plasterboard and hazardous. • Packaging re-used. • Reclaimed materials re-used. • Subcontractors encouraged to undertake best practice in waste management.	Generated waste reduced to 401 tonnes, of which 247 tonnes (685) was recycled.
Housing developer supply chain agreement with plasterboard manufacturer.	• Plasterboard segregation is a requirement on all of the housing developer's sites. • Plasterboard manufacturer collects and recycles waste plasterboard materials. • Subcontracts specify the use of the nominated plasterboard manufacturer. • Site meeting and inductions utilised to inform site workers understanding and waste management practice.	Increased recycling of plasterboard by some 4,600 tonnes per annum by third year of operation.

Table 5.7 Cost savings from lower wastage rates and waste segregation in construction averaged across a number of project analyses (WRAP 2010)

	Saving on cost of materials (A)	Reduction indisposal cost (B)	Total saving (A+B)	Cost of implementation (C)	Net Benefit (A+B)–C
			Saving as % of construction cost		
Housing	0.55	0.02	0.76	0.34	1.42
Commercial	0.44	0.16	0.06	0.17	0.43
Public	0.29	0.07	0.36	0.05	0.30
Refurbishment	0.24	0.44	0.68	0.34	0.34

(WRAP 2010)

Table 5.8 Site Waste Management – summary of net cost savings on different trade packages (each £100k value) as a result of reducing wastage rates from baseline to good practice levels on a few major components

Trade package	Construction value	Total saving (A)	Cost of implementation (B)	Net Benefit (A – B)
	(£k)		(£k)	
Brick and block wall with insulation	100	8430	1060	7370
Flooring	100	2520	1060	1460
Board insulation	100	10 620	1060	9560
Timber stud partitioning	100	2540	1060	1480

(WRAP 2010)

introduced. Furthermore this chapter has considered site waste management plans and the requirements, content and benefits of such plans.

Questions for the reader

Here follows a number of questions related to the information presented within this chapter. Try to attempt each question without reference to the chapter in order to assess how much you have learned. The answers are provided at the end of the book.

Question 1

Illustrate the ISO14001 process in a diagrammatic form.

Question 2

Section 4.2 of ISO14001 requires that an environmental policy must be put into place and be fully supported by top management. What else must top management ensure with regard to the environmental policy?

Question 3

Identify the advocated benefits of ISO 14001.

Question 4

Section 6 of the Site Waste Management Regulations 2008 stipulates the requirements of a site waste management plan. Identify the requirements laid down for a site waste management plan by Section 6 of these regulations.

References

Building Research Establishment (2006) *Developing a Strategic Approach to Construction Waste*. BRE.

ISO (2004) ISO 14001:2004: *Specification for Environmental Management Systems. International Organisation for Standardisation*. Geneva.

——(2009) Environmental Management: *The ISO 14000 family of International Standards. International Organisation for Standardisation*. Geneva. Available online at: http://www.iso.org/iso/theiso14000family_2009.pdf. Accessed 6 July 2010.

Statutory Instrument (2008), *The Site Waste Management Plans Regulations 2008*. Statutory Instrument 2008 No. 314.

WRAP (2009a), *Moving to Good and Best Practice Use of Site Waste Management Plans*, Available online at: http://www.wrap.org.uk/construction/tools_and_guidance/site_waste_management_planning/moving_to_good_and.html. Accessed 6 July 2010.

WRAP (2009b) *User Guide for WRAP Site Waste Management Plan Template v2.0*. Available online at: http://www.wrap.org.uk/construction/tools_and_guidance/site_waste_management_planning/swmp_tools_and.html. Accessed 6 July 2010.

WRAP (2009c) *Site Waste Management Plan Template v2.0*. Available online at: http://www.wrap.org.uk/construction/tools_and_guidance/site_waste_management_planning/swmp_tools_and.html. Accessed 6 July 2010.

WRAP (2010) *Reducing waste in building and civil engineering projects: Guide to cost saving and client cost saving strategies*. Available online at: http://www.wrap.org.uk/downloads/W676_Guide_to_cost_saving_and_client_cost-saving_strategies_FINAL1.a2b3d0a8.9188.pdf. Accessed 6 July 2010.

6 Developing a learning organisational culture

Project and corporate learning linked to continuous improvement provides the focus of this chapter, which fully explores the concept of developing, and the advantages of becoming, a learning organisation. The various types of learning processes are described and their importance and links to project and corporate enhancement are established for the reader and for construction-related organisations. The key issue of the control function and the vital part it plays in any improvement process, whether it is encapsulated within a quality or environmental system and at project or corporate level, is explored.

Learning outcomes

By the end of this chapter the reader will be able to demonstrate an understanding of:

- The advocated advantages of and the rationale for organisations developing a learning culture.
- The basic requirements for developing a learning organisation.
- The fundamental concepts of learning.
- How to apply the Management Functional Assessment Model.

Introduction

This chapter explores the utilisation of a self-assessment methodology based upon the key components of management for construction firms. The building blocks of the management process are defined and linked to a post-modernist learning organisational philosophy. This will enable the reader to demonstrate an understanding of the key functions of management and their impact upon construction organisational performance.

In order for construction organisations to fully engage in a continuous improvement process they must develop a learning organisational culture. The concept of organisational learning should be linked to the key functions of management; this is so because these functions control all organisational resources, procedures and systems. Therefore this chapter concentrates on establishing the fundamental rationale for the development of a learning organisational philosophy, and how its benefits may be obtained in practice for construction companies.

The Management Functional Assessment Model (MFAM) has been designed to provide a means for construction organisations to gauge the effectiveness and efficiency of their management activities linked to project and corporate requirements.

Because the model incorporates the concept of 'triple loop learning' it is possible for organisations to reflect and improve upon 'total organisational performance'.

Thus, as advocated, the model provides a means for the attainment of organisational continuous improvement.

Given that the MFAM model embraces the seven functions of management it overcomes the critical issue noted by Greising (1994):

> ... that it is too easy for managers to become overly enamoured with the procedures and mechanisms for TQM while forgetting that the point of the activity is to improve firm performance and that quality can go up but profits can go down.

Therefore it is vital that any model designed to empower improvements in all aspects of project and corporate performance must encapsulate in a holistic manner all such functions. This chapter provides the means for reflecting upon and evaluating a construction organisation's strengths, limitations and performance, and their impact on maintaining a sustainable corporate competitive advantage, built upon successful individual project performances.

It has been suggested for some years that the adoption of TQM/EFQM EM and associated methods are not widespread in smaller and medium-sized construction organisations (SMEs). The following learning philosophy is suitable for all sizes of business in both service and manufacturing sectors.

After all, as argued by Peters (1988a), ' ... if it aint broke, don't fix it' needs revision. He proposed 'If it ain't broke you just haven't looked hard enough'. The triple-loop learning incorporated in the model addresses this issue in a most simple yet effective and efficient manner for construction-related companies.

Continuous improvement can be considered an example of what many strategy theorists call 'dynamic capability' (Teece and Pisano 1994). In this approach strategic advantage is seen to come not from simple possession of assets or of particular product/market position, but from a collection of attributes which are built up over time. These provide a basis for achieving and maintaining a sustainable competitive edge in an uncertain and rapidly changing dynamic operational environment.

Normally three elements constitute dynamic capability: paths, position and processes (Tidd et al. 1997). The first two concern the amalgam of competencies that the construction organisation has accumulated and the particular position that it is able to adopt in its product/market environment. However, the third is of particular interest as it concerns the specific behavioural routines which characterise 'the way we do things in this organisation' and which describe how the construction company approaches issues of innovation, learning and improvement.

Continuous improvement represents an important element of any dynamic capability, since it offers mechanisms whereby a high proportion of the organisation can become involved in its innovation and learning processes (Bessant and Caffyn 1997, Bessant 1998, Robinson 1991, Schroeder and Robinson 1993). It corresponds to what is widely known as 'kaizen' and forms an important component of the 'lean thinking' approach adopted by many construction companies (Imai 1987). Its strategic advantage is essentially as a cluster of behavioural routines but this also explains why it offers considerable competitive potential, since these behaviour patterns take time to learn and institutionalise, and are hard to copy or transfer. The potential for continuous improvement to become an enabling mechanism in organisational learning has been advocated by Nonaka (1991) and Leonard-Barton (1995).

Human resource development in construction has previously concentrated on developing an individual's skills and knowledge related to tasks, rather than corporate learning concepts.

The need for organisations to become learning companies has been asserted by many eminent authors as a response to the increasing organisational challenges posed by rapid environmental change, discontinuity, economic uncertainty, complexity and globalisation. Indeed, one reason for the growth in popularity of the term is that it seems to capture many of the qualities deemed necessary for contemporary organisations, such as teamwork, empowerment, participation, flexibility and responsiveness. Stata (1989) argues that ' ... the rate at which individuals and organisations learn may become the only sustainable competitive advantage'. These noted components of becoming a learning organisation are impacted upon by the management functions of a construction company and these are now explored in more detail.

The seven functions of management

The task of managers may be summarised as having responsibility for forging into a holistic whole the three constituents of people, ideas and things. The attainment of this demanding task is assisted by addressing the 'Seven Functions of Management' as advocated by Fayol et al. (in the 1900s). These functions are:

- controlling
- planning

- forecasting
- organising
- motivating
- co-ordinating

Overlapping and running through the above six functions is the seventh key function of 'communication'. This is the lifeblood of any construction organisation, without which a construction manager cannot function efficiently or effectively.

An outline of the seven functions of management is presented for the reader; these outlines are provided in order to establish their importance, and their impact upon organisation quality initiatives.

Controlling

Control is concerned with the effective and efficient utilisation of resources in the attainment of previously determined objectives, contained within a specific identified plan. This plan may take many forms, e.g. a bar chart, network analysis, or a financial plan such as a project budget plan. The plan is the method that requires deployment in order to achieve predetermined set objectives. However, it should also be based upon the most efficient and effective way of completing the set task(s).

Control is exercised by the feedback of information upon actual performance when compared with the predetermined plan; therefore planning and control are very closely linked. Control is the activity which measures deviations from planned activities/objectives and further initiates effective and efficient corrective actions via a feed-forward mechanism.

In order to have both efficient and effective control the Deming Dynamic Control Loop Cycle should be employed, as depicted in *Figure 4.9* of Chapter 4. It is important that any information contained within the loop must:

- separate information according to areas of responsibility and accountability;
- present results in a consistent, readily understood and useful manner;
- represent appropriate and valid time periods for instigating effective actions;
- be available in a timely manner, enabling effective decisions to be taken;
- divert the minimum energies from corporate primary functions, consider the 'law of diminishing returns' and associated 'opportunity costs';
- demonstrate clearly any deviations from the predetermined plan and the impact of noted deviations (if at all possible).

The above can be encapsulated under the two headings of 'cycle time' – how long it takes for the information to circulate – and 'quality of information' – the level of detail encapsulated in a circulation (loop) process.

For effective and efficient control one must have short (appropriate) cycle times and the level of detail necessary (quality of information) to make valid decisions and deploy appropriate actions.

Planning

Planning is that aspect of management concerned with the particular rather than the general and is dependent upon the attainment of both reliable and accurate information. All construction managers plan; they set objectives and try to anticipate the future in order to achieve set tasks.

Construction managers determine the broad lines of operations, the strategy or general programme, choose the appropriate methods, and sometimes the materials and machines required for the most effective and efficient action. So planning relates to how, when and where work is to be carried out.

The process of planning usually refers exclusively to those operations concerned with, and the department responsible for, determining the manner in which a job/project is to be executed along with the necessary resources.

The word 'planning', in the sense of forethought, can also include such varied activities as market research, training schemes and the recording of plant locations and availability.

To be really effective, planning must be simple, flexible, balanced and based upon accurate standards of performance determined by systematic analysis of observed and recorded facts.

Planning is perhaps one of the most important tools of management, requiring intense application, precise attention to detail, imagination and a sound knowledge of technical theory, but is always a means, and not an end in itself.

In its application, construction managers should give full regard to the human needs of the organisation (covered further under the heading of motivating).

Forecasting

Forecasting or looking ahead is generally the prerogative of senior managers, although it can enter into the decision-making process at any organisational level.

The only reliable method of arriving at important policy decisions is by the adoption of a systematic approach based upon a precise diagnosis of the situation, collection and tabulation of all the facts, dispassionate consideration and prognosis, and the formulation of a logical conclusion.

Forecasting objectives can be as varied as economic forecasting, i.e. how much capital is required and which is the best source, estimating margins for tenders, the alternatives of buying and hiring plant, or the selection of appropriate personnel.

Information may be in the form of trend analysis indicated by statistical control figures, market research results or even the feedback on recruitment interview tests.

The final outcome of the forecasting process may involve things like setting tender prices, determining estimated labour requirements or even staff promotions.

Competent direction is an essential factor in efficient and effective management, and requires the qualities of broad vision, clear and incisive thinking, courage, self-confidence and a good judgement of personnel and situations.

Consideration of all factors involved is of particular importance when forecasting, as is the investigation of all possible alternatives, for the plan selected will most certainly highlight any ill-considered or uninvestigated areas and items. The planning and forecasting functions are very closely related. Fryer (1997) argues that long-range 'planning' is really 'forecasting' and the authors of this book agree.

Organising

Organising is the other aspect of management activity which is complementary to planning and concerned with the more general selection of the people and the operational methods necessary for the discharge of managerial responsibilities. Construction managers are organising when they commence putting plans into action.

The process of organising or preparing comprises:

- The definition and distribution of the responsibilities and duties of the various management and supervisory personnel forming the establishment of the enterprise;
- the recording of the types of formal relationships existing between individual appointments, the pattern of accountability and theoretical paths of contact;
- the formulation and deployment of standard procedures, preferred methods of working and operating instructions for standard techniques.

Certain guiding principles can be used by construction managers in order to determine the organisational structure; these include:

- Schedules of responsibilities, the organisational chart and standard procedures should preferably be written down and distributed so far as applicable for general reference purposes, to allow revision and to preserve continuity despite transient personnel;
- when increasing size dictates the subdivision of responsibilities, this should be determined by functional or operational specialisation;
- where possible a single head person should be responsible to the policy-forming body for the implementation of all the operations of the business;

- decentralisation of decisions should be provided by the adequate delegation of responsibility, and any limitations should be specifically noted; 'management by exception' could be employed here;
- clear lines of accountability should link the chief executive with all points of the organisation, and the integration of specialists should not interrupt the organisational lines of command and communication;
- the structure must be flexible enough to facilitate amendments when circumstances change, but since endurance is the ultimate test of an enterprise's success a formal outline and value system are necessary to assure the continuous and effective functioning of the construction company;
- a typical construction organisation does not exist, since consideration must be given to the individual characteristics and operational environments of each undertaking.

Successful construction managers divide up the total operation into individual jobs/tasks in order to be able to match them to correct personnel. However, they still have to co-ordinate them, so that one work group is not held up by another and to ensure that materials are there when required. 'The function of organising is very specific to the [construction] manager's role' (Fryer 1997).

Motivating

Many authors have argued that an organisation's most important asset, particularly in a labour-intensive industry like construction, is its people (Fryer 1997).

Motivation is a very complex function of management and there exists a wealth of published information on this topic by well-known management experts.

It is important to provide an acceptable definition of motivation and Cole (1995) provides one: 'Motivation is the term used to describe processes, both instinctive and rational, by which people seek to satisfy the basic drives, perceived needs and personal goals, which trigger human behaviour'.

It may be postulated that there are fundamentally two main types of motivation to work. One is the job as an end in itself (intrinsic satisfaction); the other is the end towards which the job provides the means (extrinsic satisfaction).

Intrinsic satisfaction

This is derived by fulfilling your own motivational needs from the job and is therefore achieved from work itself. A considerable weight of behavioural scientific research has been devoted to the pursuit of this concept.

Extrinsic satisfaction

This is deriving satisfaction of needs using work as a means to an end. It is sometimes termed the 'instrumental approach'. Work provides us with money, and money enables us to 'buy' satisfaction to a certain extent, thus pay acts as the main motivating factor.

To try and draw some conclusions from the two schools of thought it could be stated that people work for different personal reasons but, basically, they fall into two categories: extrinsic and intrinsic. How can a construction manager motivate a workforce which is most likely to be a mixture of the two categories? For the extrinsic workers it would appear that financial stimulus is the only means of motivation. This tactic has been tried in the form of incentive payments and bonus systems and indeed these usually do work for this type of employee. The opportunity to earn more is taken up and production increases (though this tends to be only a short-term phenomenon). What can managers do about their intrinsically motivated employees, who may never be motivated by money alone?

Intrinsic workers require motivation from the task and in order to achieve a desirable state of high morale it is necessary to:

- Arouse interest by keeping everyone informed of proposed developments and the progress of activities/set tasks.
- Foster enthusiasm by assisting in the attainment of legitimate personal and social satisfaction.
- Develop harmony and a sense of participation by engaging in joint consultation processes.
- Enlist co-operation by providing reasonable continuity of employment and security for their future.
- Secure loyalty by showing fairness and being consistent in the allocation of duties, distribution of rewards and administration of discipline etc.
- Promote keenness by fostering a sense of competition (where appropriate) and group or personal achievement.
- Encourage self-discipline by developing a sense of responsibility and the enjoyment of trust.
- Inspire confidence and respect by fair judgement and impartial dealings with subordinates.
- Ensure acceptance of the necessary rules and regulations by inspiring a sense of duty and a responsibility for the affairs of the construction organisation and individuals.
- Assist ambition by the encouragement and the affording of opportunities for individual development.
- Prevent frustration by providing a sympathetic and effective outlet for grievances and misunderstandings.
- Provide timely feedback on performance.

Motivation is a vitally important concern to both employees and managers within an organisation. Its importance arises from the simple but powerful truth that poorly motivated people are likely to perform poorly at work and gain little satisfaction from their job.

(Naoum 2001)

Co-ordinating

This is the linking together of the various members to constitute a practical ensemble and the balancing of resources and activities to ensure as far as is practicable the complete harmony of processes and performance.

The main aims of the co-ordination function are first to prevent the separation of activities into water-tight compartments as a sequence of specialisation, and second to preserve a recognisable unity throughout the enterprise. Or in other words to ensure a truly holistic approach to all construction organisational activities, bounded by a set of common goals.

A tendency for organisational activities to separate into individual functions increases with the size of the company, therefore keeping a functional team together may become a vital task for a construction manager.

Deliberate co-ordination of management may require specified activities such as regular meetings to integrate ideas and actions, the establishment of an additional effective communication system, and possibly for the attainment of greater clarity a pictorial presentation of responsibilities to assist co-operation between individuals and teams.

The performance of successful co-ordination requires:

- early introduction of the function;
- direct personal contact with all parties concerned;
- a reciprocal activity by the personnel being co-ordinated;
- continuous operation of the function and the monitoring of its effectiveness.

Co-ordination is achieved in the main by the efforts and skill of the individual construction manager with due regard to the overriding human factors involved. It is the assessment and utilisation of core skills in the best interests of the organisation and its employees, with particular emphasis being placed on the harmony of major resource elements.

A construction manager has to co-ordinate a very diverse range of resources; these are the 'Five Ms': materials, manpower, management, money and machines. They must also integrate the work of subcontracts, all within a specific time frame and directed towards the achievement of set project/corporate tasks.

Communication

Communications are the means employed by executives to pass on their plans and instructions for action and by managers to make known their objectives/requirements, and to inspire the necessary efforts, and by supervisors to co-ordinate activities and control operations.

Communication is a means of achieving contact between departments and individuals, and a channel for the distribution of knowledge; these are obvious fundamental activities of construction management. However, equally important aspects of communicating are not always fully appreciated by construction managers and these incorporate the following:

Good communications promote a better understanding by describing what is being done and why, and permit free expression of suggestions by all levels of personnel. This encourages a sense of participation and prevents friction and misunderstandings, thus contributing to the achievement and maintenance of healthy staff morale.

Although the process of communicating is the indispensable 'tool' of management or supervision, nevertheless the ability to convey messages clearly, vividly and convincingly, by either speech or writing, is the key to the exercise of power. The spoken word is more infectious and particularly useful for short-term persuasion, while the written word is more permanent and hence usually more suitable in the long term.

To be effective both arts necessitate the possession of accurate facts, the use of simple, precise language, and the fluency of expression developed via constant practice by construction managers.

It should be remembered that 'Many [construction] organisational problems are caused by communication failure. Breakdowns occur because of faulty transmission and reception of messages and because people put their own interpretation on what they see and hear' (Fryer 1997).

Communication with employees can be defined as the passing on and receiving of signals from one human being to another. (Of course computers can now serve this purpose.) The purposes of communication are threefold:

- to increase knowledge and/or understanding; the construction manager may not want their workforce to change their behaviour;
- to influence or change attitudes; although direct verbal communication may be designed with a view to changing attitudes it is unlikely to do so;
- to instigate or influence action or behaviour. Ultimately all communication, particularly in the workplace, is perhaps geared to this end.

Although construction managers may seek to increase knowledge or change attitudes they will only have proof of having done so if behaviour of the recipient/teams changes as a result.

Overview of functions

Construction managers' jobs are demanding, complex and varied; there are a set of common features in the role of these managers, but individual jobs do differ considerably. Thus the amount of time and the way construction mangers perform the previously noted functions of management depend very much on their abilities and personal motivation, and the motivation of their teams and team members.

> In the 1990's there was a growing interest in empowerment, a process which shifts some of the power from the managers to employees, individually and as self managed teams. Employees, being closer to the workforce and having a superior knowledge (in some cases) of the work and its environment, are often in a better position to make decisions; empowerment gives them the opportunity to use and develop their talents more fully.
>
> (Fryer 1997)

The above noted, there is a fine line between management delegation and abdication. Empowerment should be a shared joint experience/venture between construction managers and their employees. It is not and should not be an excuse to abdicate a construction manager's responsibility. Thus the activities/functions described in this section are still valid if a construction organisation is to be competitive and continually improve.

Construction managers can also act as catalysts for change within the context of organisational culture.

> The culture of an organisation is its customary and traditional way of thinking and of doing things, which is shared to a greater or lesser degree by all its members, and which new members must learn, and at least partially accept, in order to be accepted ... [it] covers a wide range of behaviour: the methods of production; job skills and technical knowledge; attitudes towards discipline and punishment; the customs and habits of managerial behaviour; the objectives of the concern; its way of doing business; the methods of payment; the values placed on different types of work ... and the less conscious conventions and taboos.
>
> (Jacques 1952, cited by McCabe 2001)

Senior managers can clearly affect organisational culture and organisational culture can assist or obstruct organisational change-management processes; when considering organisational change processes senior construction managers must demonstrate that they are 'capable of driving cultural change through various organisational levels [and they must also have]

- Leadership skills
- Motivation skills
- Skill in dealing with resistance
- Skill in recognising different approaches, values and norms [impacting on culture] ...

The sort of person appointed to manage the implementation of culture change will need to be someone who, as well as being absolutely competent in technical issues surrounding the subject, possesses abundant confidence and human relation skill.

(McCabe 2001)

The critical importance of the seven functions of management has been outlined. It is clear that in order to manage quality and improvement processes a construction company must understand the seven functions of management, and how it is performing in relation to them. This performance measurement is encapsulated within the Management Functional Assessment Model (MFAM).

Having outlined the seven functions of management we must also acknowledge that traditional approaches to management are inadequate for keeping pace with changes in construction-related dynamic operational environments (Bounds et al. 1994).

This being the case, a new approach for considering organisations and their environmental interaction has been developed and this new philosophy is termed 'postmodernism'. As noted by Jackson and Carpenter (2000) 'That it [postmodernism] has relevance to the understanding of [construction] organisational behaviour is not in doubt.' Thus the importance of a postmodernist philosophy would need to be incorporated into any change process/improvement model in order to improve on the effectiveness of managers and hence corporate and project quality-management activities.

In the next section we shall explore postmodernism in a little more detail. It is important that construction companies have an understanding that this approach can positively impact upon corporate and project performance and hence the quality of their product/service provision.

Question 1 for the reader

Define the terms 'Intrinsic and Extrinsic Motivation'. Answers are provided at the end of the book.

Postmodernist philosophy

Successful construction companies are those that have changed their business processes due to re-evaluating their understanding of how business activities should be organised/conducted. They may have had to abandon previous organisational and operational procedures and created new more appropriate ones. It is likely that previous procedures were based upon assumptions related to technology, people and organisational goals which may no longer hold true and hence lack validity within their changed operational environments.

Modernist theory assumes that change is a linear process and can be managed in an incremental way with distinctive points of conception and completion. In essence it is a belief in a simple cause-and-effect relationship. However, a more realistic view of construction operational environments rejects the notion of linearity. Postmodernist organisations realise that change develops in many directions and the world is best understood in terms of disorder and unpredictability. A postmodern philosophy recognises the need for versatility and the emphasis is placed on organisational flexibility and quick response.

Within the following section a more detailed analysis of the differences has been undertaken and the advantages of postmodernism established for construction companies/managers.

Modernism *versus* postmodernism

In times of static or limited dynamic environmental change a modernist organisational structure can cope with changes reasonably well. However, when the environment becomes more dynamic and complex the modernist-structured organisation finds it difficult to cope with the implications of change.

Passmore (1994) noted that 'Most of us are born with a good deal of flexibility; it's a helpful trait that allows our species to adapt to the wide range of habits and circumstances we encounter. But the process of growing up in a hierarchical world [as are some construction companies] teaches us to become inflexible.'

Passmore is arguing that people are inherently capable of dealing with change and it is the bureaucratic systems they work within that stifle their intrinsic flexibility. Most modernistic construction firms would fall under this umbrella.

Early authors on this subject, such as Weber, claimed that

> Modern business enterprises are structured as 'rational-legal' hierarchical and bureaucratic systems characterised by standardised operating procedures, regulations, performance standards and 'rational' decision-making processes [not suitable for deploying TQM/EFQM EM] that are based upon technical and professional expertise.
>
> (Weber 1908)

The above is now being contested by various authors, such as Morris and Brandon, who suggest that there has been a paradigm shift in the way construction organisations view themselves and their operational environments. After all,

> When the business world undergoes change, only those companies that react quickly will prosper. The ability to react requires considerable

flexibility and openness to new ideas and approaches [most suitable for deploying TQM/EFQM EM]. In creating this foundation the basic assumptions of the business must be re-examined.

(Morris and Brandon 1993)

The above-noted paradigm shift is apparent in the postmodern organisation.

Structure of relationships

Within modernist organisations there exist very simple structure or boundary relationships. Linkages are achieved through formal rules and procedures, and relationships between different groups are formalised. In comparison the postmodernist organisation possesses little distinctiveness of roles, and boundaries are blurred. The emphasis is placed upon creating teams with positive productive relationships, all directed at increasing the organisation's ability to cope with change, because this is necessary for them to be creative. Majaro (1992) points out that making the change to a postmodernist organisation ' ... is easier said than done ... ' and that 'One of the most difficult challenges to any organisation is the process of changing a climate or corporate attitudes'. It is undoubtedly a difficult change process for any construction organisation to undergo, but the benefits are well worth the effort.

Hierarchy

Modern organisations have a very distinct hierarchy, with clearly defined leadership roles fixed by legitimacy and tradition; there are leaders and followers. Contrasted with this is the postmodern organisation, where normal hierarchy does not exist and staff act according to agreed areas of expertise. The term for this approach is 'hecterarchy', in which very high levels of fluidity are maintained. The high level of fluidity is a basic necessity for construction organisations because 'Too much is changing for anyone to be complacent' (Peters 1988b). As construction organisations move to areas of increased complexity of service, there is a requirement to implement increasing hecterarchic ways of operating and construction companies/managers must be cognisant of this fact.

Mechanistic versus holographic

In the modern construction organisation the relationship between tasks is of a mechanistic nature and there is a high degree of linear relationship between tasks. However, within the postmodern organisation high levels of group work exist, each with a correspondingly high level of autonomy. The overriding linking force binding these empowered groups together is that of organisational culture. This form more readily suits the reality of today's changing business environment because organisations, and the markets they operate in, are messy and not linear. Building a shared culture and conception of the world takes a great deal of time and effort on the part of construction firms. Traditionally some less enlightened construction companies have had a culture based upon mistrust, and the use of frequent sanctions by managers and senior managers.

Determinacy versus interdeterminacy

The modernist construction organisation conducts all matters in a determinate manner, where a high degree of emphasis is placed on imposed stability, control and discipline. This assumes that a construction firm can exercise a high degree of control over their environment, but this is not the case in reality.

Postmodern organisations conduct matters in such a way that they emphasise indeterminacy. This is an acknowledgement that the environment is highly unpredictable and uncertain and it has different values from the modernistic organisation; for example, flexibility and innovation are highly prized traits.

> Flexible people are open minded, willing to take reasonable risks, self-confident, concerned and interested in learning. They are creative and willing to experiment with new behaviours in order to make better choices about what works for them and the organisation. They possess basic skills that allow them to adapt readily to new circumstances, and they view themselves as able to make the best of opportunities that come their way.
>
> (Passmore 1994)

This in essence is the postmodernistic organisation and one construction firms/managers should seek to emulate.

Causality

A major difference between modern and postmodern construction organisations is that modernistic ones view causality as having a linear relationship. They view every element of organisational life as having a cause and effect and consequently they manage their organisation in this light. However, postmodernistic ones think in terms of a 'circle'. They are encouraged to look for complexity and the interconnection of cause and effect. This demands a high level of staff participation and makes good management sense. The rationale for participation has been stated as follows: 'When subordinates are consulted about and contribute to the change process [for example quality improvement processes] many benefits accrue' (Sayles 1989).

A more enlightened view of corporate strategy is that it is in fact 'emergent' and not linear, and hence construction companies can better cope with this if they have adopted a postmodernist approach to business and people.

Morphostatic versus morphogenic

Morphostatic processes are defined as those that support or preserve the present mode of operation and include formal and informal control systems, with the emphasis being placed on procedures. This approach is not at all conducive to learning, or seeking to improve corporate/project performance. A more enlightened approach is adopted by the postmodern construction organisation where a morphogenic culture exists. Morphogenic processes are those that tend to allow for change and development and the exciting nature of change is always advocated and accepted. This type of construction organisation encourages staff to be proactive in all areas and functions, and hence always seeks to learn and improve corporate/project performance.

Customer focus

Construction companies have to be open to ideas related to organisational improvement, thus they should be learning organisations (this issue is expanded upon in the next section of this chapter). They must consider new management tools and procurement systems that have proved to be most advantageous in manufacturing industries. These tools and methods of operation can assist in differentiating a company from its competitors. The application of a postmodernist approach to managing construction companies could provide the following advantages:

- companies are more flexible and therefore better able to cope with the demands of a changing and challenging operational environment;
- the attainment of enhanced teamwork and participation at all levels of the company leading to improved communications;
- organisational culture is highly motivated and proactive, leading to increased participation and productivity;
- enhanced corporate innovation, and learning becomes embedded;
- improved product/service quality becomes the norm;
- a greater market awareness and thus enhanced stakeholder satisfaction.

The identified characteristics of the postmodernistic company are essential for a construction organisation to be able to operate both efficiently and effectively in a dynamic and turbulent operational/economic environment.

Construction firms require variety in their approach and hierarchical authoritarian organisations are poorly equipped to provide such variety. Only construction organisations based on the postmodern model, with vastly reduced bureaucratic control, a rich array of horizontal communication channels, and in which personnel are given a substantial share of authority to make choices and to develop new ideas, are likely to survive.

With the above in mind the Management Functional Assessment Model (MFAM) has been developed. Due to its requirement for continually evaluating the management functions and activities it truly enables a move towards a postmodernist philosophy to be achieved, set within a learning organisational culture.

For example, as indicated in *Table 6.1*, under criteria 2 'Organising', can be found the sub criteria of

2.1 Creating the correct organisational structure.
2.3 Creating a self-learning organisational culture.
2.4 Developing a value system based upon enhancing performance.

These are not mutually exclusive from the requirements of postmodernism (see *Table 6.1* for further elements of postmodernism encapsulated in the MFAM).

Question 2 for the reader

Define the advantages of adopting a postmodernist philosophy for a construction company. An answer is provided at the end of the book.

Definition of a learning organisation

There is no clear consensus as to what constitutes a learning organisation, and a multitude of definitions abound. These range from aspirational type definitions of organisations, ' ... where people continually expand their capacity to create the results they truly desire, where new and expansive patterns of thinking are nurtured, where collective aspiration is set free, [truly in line with a postmodernist philosophy] and where people are continually learning how to learn together' (Senge 1990), to more normative definitions such as that espoused by Garvin: 'A learning [construction] organisation is an organisation skilled at creating, acquiring, and transferring knowledge, and at modifying its behaviour to reflect new knowledge and insights' (Garvin 1993). Garvin's views have been fully encapsulated within the MFAM.

Nyhan et al. (2004) suggested that ' ... the prescriptive and simplistic formula based view of the learning organisation does nothing more than discredit the concept'. In their opinion, becoming a learning organisation involves more than simply applying a formula; each individual organisation needs to ' ... devise its own unique theory based on its own distinctive practice' (Nyhan 2004). Once again the advocated MFAM empowers this approach.

Table 6.1 Marking criteria for MFAM respondents

Criteria	Assessment category	Max	Score
1.Forecasting/Planning	1.1 Setting the objective and strategic planning process in motion	4	
	1.2 Gathering and analysing information related to clients and markets	4	
	1.3 Detailing business processes	4	
	1.4 Gathering and analysing information related to competitors and benchmarking	4	
	1.5 Resources planning	4	
	Criterion total	20	
2. Organising	2.1 Creating the correct organisational structure	4	
	2.2 Establishing appropriate authority and responsibility for all personnel	4	
	2.3 Creating a self-learning organisational culture	4	
	2.4 Developing a value system based on enhancing performance	4	
	2.5 Deployment of new technology linked to corporate enhancement	4	
	Criterion total	20	
3. Motivating	3.1 Developing a co-operative culture based upon stakeholder satisfaction	4	
	3.2 Ensuring staff have the skills, competences and resources to perform set tasks	4	
	3.3 A consideration of personnel needs linked to self-actualisation	4	
	3.4 Engagement in processes, increase areas of responsibility and self-monitoring	4	
	3.5 Results satisfaction – feedback on performance in a timely manner	4	
	Criterion total	20	

Table 6.1 (continued)

Criteria	Assessment category	Max	Score
4. Controlling	4.1 A monitoring system for each key stage of business-process	4	
	4.2 Measuring performance levels	4	
	4.3 Determining customer satisfaction levels	4	
	4.4 Determining the efficiency and effectiveness of resource utilisation	4	
	4.5 Conducting a comparative analysis between set targets and actual results, leading to appropriate actions	4	
	Criterion total	20	
5. Co-ordinating	5.1 Unity of all other functions	4	
	5.2 Establishing effective internal communications	4	
	5.3 Developing a conflict-solving culture	4	
	5.4 Updating deviations: revision and possible reco-ordination of other resources	4	
	5.5 Ensure effective information management	4	
	Criterion total	20	

Historical development

The concept of the learning organisation has been around for quite some time; Burns and Stalker (1961) published their theory of mechanistic and organic systems following lengthy studies of a large number of Scottish electronics companies operating in increasingly competitive and innovative technological markets during the 1950s.

The 1980s was a decade of immense upheaval for many large corporations who increasingly found ' ... their success eroded or destroyed by the tides of technological, demographic, and regulatory change and order of magnitude productivity and quality gains made by non-traditional competitors' (Hamel and Prahalad 1994).

A new wave of literature appertaining to learning organisations emerged during this period, heavily influenced by organisational learning and action learning theories such as those developed by Revans (1983). Much of the work from this period recognised that corporate survival in the new global competitive environment was dependent on an organisation's ability to learn faster than its competitors, and that this ability may be the organisation's only form of sustainable competitive advantage (De Geus 1988).

Nonaka (1991) recognised that in a global economy typified by shifting markets and technological proliferation, successful construction companies will have to ' ... consistently create new knowledge, disseminate it widely throughout the organisation, and quickly embody it in new technologies and products'.

However, it was Senge's seminal text, *The Fifth Discipline* (Senge 1990), that really popularised the concept of the learning organisation. Senge described five vital dimensions or 'disciplines' that he considered to be essential for an organisation to become a truly learning company, they are: team learning, personal mastery, managing mental models, shared vision and systems thinking.

The fifth discipline, systems thinking, was seen as the integrating discipline that unites the organisation, individual and total environment, based on a conceptual framework that describes a system as a set of interrelated subsystems. Senge (1990) proposes that it is the relationship between these subsystems that ultimately influences the functioning of the whole. This concept is not mutually exclusive from the concept of the integrative nature of the seven functions of management and their impact upon construction organisational performance at corporate and project level.

Senge's work, however, has been criticised for paying insufficient attention to knowledge management systems, the structures of the organisation and their implication as a resource to learning (Sun and Scott 2003), whilst Garvin (1993) considers Senge's model as too ethereal, and lacking a 'framework for action'.

The nature of a learning organisation culture

Much of the discussion in the management literature is clearly written from the perspective that the learning organisation can be designed and managed effectively to produce positive outcomes for the organisation. Many commentators have attempted to specify what the learning organisation culture should consist of. And although numerous authors (e.g. Garvin 1993, Senge 1990) have considered the notion of a learning organisation culture, there is no widely accepted theory or view on this issue. Some have identified specific aspects of a learning organisation culture such as entrepreneurship and risk taking. Indeed the literature on learning culture characteristics is extremely broad, drawing on work from sociology, psychology and anthropology as well as business disciplines, which perhaps makes the task of formulating such a theory a monumental one.

The nature of the learning process

Pedlar et al. (1998) points out that the aforementioned writings (Senge 1990) take such a wide view of the structures in which an organisation does or needs to learn, that the idea of learning becomes lost. They cite four questions which are of relevance here: What is learning? Are there types or levels of learning in organisations and are they recognised? What are the different levels of learning? How does an organisation facilitate or inhibit the learning process? Construction companies must fully consider these important questions.

They suggest that there is a need to hold on to the idea of the learning organisation as a direction, while organisational learning, which is a fundamental component of the learning organisation, is seen as a heuristic device to explain or quantify learning activities. This would seem to suggest that an emphasis be put on understanding how learning is defined, acquired and used at the individual and organisational level.

Rarely do construction firms have an understanding of what it is they are measuring and, when they do, they may be only measuring activities as part of an organisational control system. However, a major challenge for them will be to develop valid measures of learning outcomes specifically to assess whether they have actually learned, demonstrated by changed behaviour and project/corporate improvement.

Learning cultures

The concept of culture itself is intangible and the notion of a learning culture is perhaps easier to experience than describe. There is evidence, however, to suggest that an organisation's culture may facilitate or inhibit learning depending on its characteristics (Argyris et al. 1978). They suggest that an organisation's defence routines may be both anti-learning and over-protective. They further argue that such patterns of behaviour may become so embedded in the culture that they are rarely questioned or challenged.

Cummings (2005) emphasises that it is important for a firm's culture to be supportive because it is difficult to develop and sustain appropriate learning behaviours if the corresponding organisational values are not in place (this requires a supportive paradigm, e.g. morphogenic), and similarly values are difficult to sustain if the appropriate incentives and examples do not exist. This suggests a synergistic relationship between the elements of culture and learning activities within the organisation systems which support the learning organisation. He further suggests that culture often embodies an accumulation of prior learning, based on earlier success.

Underpinning cultural values

Building construction learning organisations is, in effect, an attempt to manage the culture of the said organisations and it requires specific attention to some key cultural values if it is to be a successful undertaking. For example they need to address:

- *Celebration of success.* If excellence is to be pursued with vigour and commitment, its attainment must be valued within the organisational culture.
- *Absence of complacency.* Learning organisations reject the adage 'if it ain't broke don't fix it'; they are searching constantly for new ways of delivering products and services. Thus innovation and change are valued and promoted within the organisation.
- *Tolerance of mistakes.* Learning from failure is a prerequisite for progressive organisations. This in turn requires a culture that accepts the positive spin-offs from errors, rather than one that just seeks to allocate blame. However, this does not imply a tolerance of routinely poor or mediocre performance from which no lessons are learned.
- *Belief in human potential.* It is people that drive success in organisations, using their creativity, energy and innovation. Therefore the culture within a learning organisation values people, and fosters their professional and personal development.
- *Recognition of tacit knowledge.* Learning organisations recognise that those individuals closest to processes have the best and most intimate knowledge of their potential and flaws. Therefore, the learning culture values tacit knowledge and shows a belief in empowerment (the systematic enlargement of discretion, responsibility and competence).
- *Openness.* Because learning organisations try to foster a systems view, sharing knowledge throughout the organisation is one key to developing learning capacity. 'Knowledge mobility' emphasises informal channels and personal contacts over written reporting procedures. Cross-disciplinary and multifunction teams, staff rotations, on-site inspections, and experiential learning are essential components of this informal exchange (a postmodern approach).
- *Trust.* For individuals to give of their best, take risks and develop their competencies, there must be demonstrable trust.

Örtenblad's (2004) 'learning structure' model builds on this idea by describing a decentralised, flatter organisational structure that is team based, with learning depicted as an input, and flexibility as an output.

Providing the correct corporate environment

Garvin (1993) suggests that a learning organisation is one that fosters ' … an environment that is conducive to learning'. He purports that in order for employees to learn, they need ' … time for reflection and analysis, to think about strategic plans, dissect customer needs, assess current work systems, and invent new products'. This highlights an important prerequisite for implementing any new initiative: the provision of adequate resources, particularly those of time and funding, which are not mutually exclusive.

However, time and money alone will not create the required climate for learning. Ho (1999) proposes that the learning organisation provides an environment where ' … people are excited in trying out new ideas and recognise that failure is an important part of success'.

Love (2004) underpins this view, describing an atmosphere where ' … experimenting with new approaches is encouraged [not common in construction] and errors are not perceived as failures'. These traits, when viewed in the context of an organisational environment typified by ever-increasing complexity and uncertainty (Malhotra 1996), clearly point towards a requirement for a morphogenic culture utilising processes that ' … allow for change and development … [and where] the exciting nature of change is always highlighted' (Griffith and Watson 2004).

Knowledge management

In the late 1980s, Pedler et al. (1988) recognised the importance of utilising information technology to ' … informate as well as automate … [in order to] seek information for individual and collective learning.' However, Lobermans has asserted that a 'corporate architecture' needs to be in place to facilitate learning and to ' … create knowledge sharing and dissemination mechanisms across the organisation' and that the capture and systemisation of knowledge is a prerequisite to being a learning organisation (Lobermans 2002). The growing number of organisations utilising intranets and 'lessons learned' databases gives some indication of the perceived value of knowledge management systems to the construction industry.

However, recent research into cross-project learning led Newell to conclude that ' … there is accumulating evidence that the medium of capture and transfer through ICT such as databases and corporate intranets is limited in terms of how far such technology can actually facilitate knowledge sharing' (Newell 2004). Newell's study also found that where transfer of learning had occurred, it had depended far more on social networks and a process of dialogue than on ICT.

These findings concur with the view of Nonaka (1991) in that the key to construction organisations gleaning greater knowledge is through facilitating:

- the sharing of tacit knowledge through socialisation;
- the collation of discreet pieces of explicit knowledge to create new knowledge;
- the conversion of tacit knowledge into explicit knowledge, i.e. externalising what individuals know;
- the conversion of explicit knowledge into tacit knowledge, i.e. internalising explicit knowledge.

The key features of construction learning organisations relate less to the ways in which organisations are structured and more to the ways in which people within the organisation think about the nature of, and the relationships between, the outside world, their colleagues and themselves.

Of course the key focus for all organisational activities should be the satisfaction of the client, and the learning culture is directed to this end product.

Crucially, learning construction organisations do not focus exclusively on correcting problems or even on acquiring new knowledge, understanding or skills. They aim instead for more fundamental shifts in organisational paradigms and try to encourage the development of learning capacity at all levels.

Mental models

Senge's (1990) discipline of managing mental models recognises that ' ... new insights fail to get put into practice because they conflict with deeply held internal images of how the world works, images that limit us to familiar ways of thinking and acting'.

Argyris and Schön (1974) opined that people are often unaware that the mental models that inform their actions are often not founded in the beliefs that they explicitly advocate, leading to a contradiction between their espoused theory and their theory in practice. In order for people to manage their behaviour more effectively, they suggest the use of double-loop and even triple-loop learning, in order to develop congruence between theory and deployed practice.

The double-loop learning advocated by Argyris and Schön is fundamentally what Senge was referring to when he suggested that mental models should be brought to the surface and reflected on by ' ... balancing advocacy and inquiry', a process he describes as being ' ... open to disconfirming data as well as confirming data – because we are genuinely interested in finding flaws in our views' (Senge 1990).

This contemplative approach is necessary in order for construction organisations to escape what Shukla calls 'the success trap' (1997). He describes how successful construction companies try to replicate their achievements by formalising their effective practices and procedures, standardising their products and services and investing in tried and tested technologies.

This single-loop approach to learning results in the construction firm becoming less sensitive to competitive demands; they lose touch with their environment and, as Shukla explains, ' ... their past learning becomes a hindrance in the way of the necessity of new learning; they must "unlearn" to learn' (Shukla 1997).

Hamel and Prahalad (1994) use the term 'frame' in place of mental model, proposing that 'Although each individual in a [construction] company may see the world somewhat differently, managerial frames within an organisation are typically more alike than different', and 'Almost by definition, in any large organisation there is a dominant managerial frame that defines the corporate canon'.

The suggestion that there can be an institutional model echoes the view espoused by De Geus who sees the mental models of each learner as 'a building block of the institutional mental model' (De Geus 1988, Cummings 2005).

Single-, double- and triple-loop learning explained in more detail

Argyris and Schön (1974) first developed the idea that there are two basic types of organisational learning, 'single loop' and 'double loop'. Single-loop, as noted, is learning where organisations respond to changes in their internal and external environments by detecting and correcting errors in order to ' ... maintain the central features of the organisational norms' (Barlow and Jashapara 1998). Argyris (1996, cited Dahlgaard 2004) when considering learning within an organisational context, suggests that an error is any mismatch between the intention and what actually happens (the results). However, he further argues that discovering errors is not really learning and that learning only occurs when the discovery or insight is followed by an action. From this viewpoint, learning inevitably involves the taking of some action.

Single-loop learning

Single-loop learning assumes that problems and their solutions are close to each other in time and space (although they often aren't). In this form of learning, we are primarily considering our actions. Small changes are made to specific practices or behaviours, based on what has or has not worked in the past. This involves doing things better without necessarily examining or challenging our underlying beliefs and assumptions. The goal is improvements and fixes that often take the form of procedures or rules. Single-loop learning leads to making minor fixes or adjustments.

It could be argued that incremental, imitative learning methods such as benchmarking and best practice are examples of single loop learning. Within what Argyris described as single-loop learning, decisions are based solely on observations while in double-loop learning decisions are based on both observation and thinking.

Learning hasn't really taken place until it's reflected in changed behaviours, skills and attitudes (Stata 1989).

Double-loop learning

Double-loop learning involves a more demanding approach to learning, where an organisation's norms, policies, assumptions and past actions are critically examined in order to inform new strategies for learning (Argyris and Schön 1974). Inevitably, such introspective organisational analysis may bring about conflict; Love (2004) maintains that 'Frequently organisational conflict is a correlate of double loop learning in as much as the status quo is challenged [as an organisation moves towards a more morphogenic culture].'

In summary it can be stated that in single-loop learning people's decisions are based solely upon observations, while in double-loop learning decisions are based on both observation and thinking.

Double-loop learning leads to insights about why a solution works. In this form of learning we are primarily considering our actions in the framework of our operating assumptions. This is the level of process analysis where people become observers of themselves, asking 'What is going on here? What are the patterns?' And we require this insight in order to understand the patterns. Double-loop learning works with major fixes or changes, like redesigning an organisational function or structure.

In triple-loop learning a reflection phase is incorporated to support or improve the thinking phase and hence to improve the decision-making process. 'Thus both double and triple loop learning can be considered as generative learning, while single loop learning can be considered an adaptive learning' (Dahlgaard 2004).

The developed MFAM employs the concept of triple-loop learning.

Triple-loop learning

Triple-loop learning goes beyond insight and patterns and the result creates a shift in understanding the corporate context, or point of view, where new commitments and ways of learning are produced. This form of learning challenges us to understand how problems and solutions are related, even when separated widely by time and space. It also challenges firms to understand how previous actions created the conditions that led to current problems. The relationship between organisational structure and behaviour is fundamentally changed because the organisation learns how to learn.

Summary

The concept of the learning organisation has evolved as a response to a rapidly changing, dynamic business environment, which is constantly in flux.

The idea, then, of a fluid, flatter, less hierarchical organisational structure that offers less resistance to the seepage of knowledge through the organisation appears to have credence.

An organisational structure only provides the skeleton of the learning organisation; a capillary system is necessary in order to transfer knowledge around the organisation at all levels. It does seem that most knowledge-management strategies focus solely on the electronic collation of information, failing to take account of how different types of knowledge are internalised and externalised via the use of already existing social networks.

There also appears to be a degree of consensus that a 'learning climate' has to be created, where individuals feel free to experiment with new ways of doing things. However, this requires a blame-free culture where mistakes, instead of being hidden, are acknowledged and learned from. Changing organisational culture requires a well-planned change-management strategy to be developed and this has to be initiated and supported by senior management for it to have any chance of success. It does seem that the utilisation of 'mental models' by construction companies inhibits the implementation of new concepts and most models are based on replicating previously effective practices. The models, though individually held, collectively form and reinforce the organisational model, which is focused on maintaining the status quo. The MFAM is designed to challenge the status quo, with a view to obtaining organisational improvement.

The idea of surfacing mental models (Senge 1990) seems closely aligned with the concept of double-loop learning (Argyris and Schön 1978). The introspective organisational analysis associated with both concepts is a quantum leap away from the morphostatic culture (Griffith and Watson 2004) prevalent in many organisations, and may prove to be one of the most difficult learning organisation characteristics to attain. There is clearly a need to make a change-management strategy an integral part of any generic implementational model.

Other key characteristics that typify a learning organisation are:

- having in place a strategy for creating, acquiring and disseminating knowledge;
- collective aspiration (a shared vision);
- an emphasis on continuous learning leading to continuous improvement;
- a holistic, 'systems thinking' approach to learning that recognises the interrelatedness of the organisation, the individual and its external environments;
- a tolerance of some experimentation by people.

Several problematic issues may prevent a construction company from successfully implementing learning organisation concepts. For example, organisational structures geared towards stability rather than change as identified by Johnson and Scholes (2002) are noted as an unsuitable framework upon which to found aspirations to become a learning organisation.

A lack of senior management support, resulting in failure to provide adequate resources, particularly in respect of allowing employees 'time to think', will also lead to failure.

The above has been highlighted in order to emphasise the importance of triple-loop learning being incorporated into any model designed to improve the effectiveness of management functions. The MFAM does indeed incorporate triple-loop learning.

Question 3 for the reader

Define the terms single loop, double loop and triple loop, with regard to organisational learning. Answers are provided at the end of the book.

Management Functional Assessment Model (MFAM)

The behaviour of an organisation's leaders should create a clarity and unity of purpose within the company and an environment in which its personnel can learn and improve. A truly empowered organisation employs both a top-down and a bottom-up approach to managing and performing its organisational activities.

Construction companies perform more effectively and efficiently when all interrelated activities are understood and systematically managed, and decisions concerning current operations and planned improvements are made using reliable information that includes both stakeholder perceptions and expectations.

Corporate performance is maximised when it is based on the management and sharing of knowledge within a culture of continuous triple-loop learning, innovation and improvement. The penalties of failure must not outweigh the rewards of success, or this will undermine any attempt to encourage a culture of innovation and risk taking set within the context of a learning organisation culture.

A construction company works more effectively and efficiently when it has mutually beneficial relationships built on trust and the sharing of knowledge and integration with its partners. Therefore leadership and culture are vital components of a continuous improvement learning process. This fact has been recognised by the European Foundation for Quality Management in the development of its 'RADAR' concept as noted in Chapter 4 of this text book. RADAR is indicated in *Figure 6.1* (European Foundation for Quality Management 1999).

Obtaining a sustainable competitive advantage via MFAM deployment

For a construction company to attain a sustainable competitive advantage they require a competitive-orientated management system. The advocated management system should embrace the issues previously described within

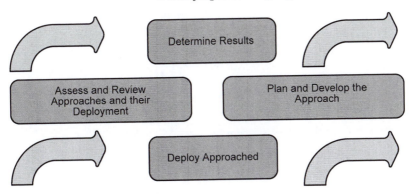

Figure 6.1 The criteria underpinning the RADAR concept

this chapter, namely the seven functions of management, a morphogenic philosophy and performing as a learning organisation utilising triple-loop learning, all as part of a holistic management process.

The system should also address key concepts such as leadership, personnel and development. However, the system must also fully address the needs of a company's stakeholders.

The MFAM is based upon previous works including those of Deming, Baldrige and the European Foundation for Quality Management. However, it is original and when deployed provides an effective link between all organisational activities set within a framework of corporate profitability and stakeholder satisfaction.

This section explores the model's constituent parts and relates them to the process of gathering data on organisational performance and requirements, focussed upon attaining/maintaining a competitive advantage. The key concept of 'RADAR', which forms part of the European Foundation for Quality Management Excellence Model, is embedded within the MFAM, thus embedding the triple-loop learning concept.

Corporate excellence is measured by an organisation's ability to both achieve and sustain a competitive advantage through satisfying its stakeholders. This can only be achieved by the efficient and effective utilisation of all corporate resources which include the Five Ms. The Five Ms have to be treated as a holistic whole and the MFAM provides a means for setting corporate objectives that are linked with stakeholder expectations and needs. The advocated model enables construction firms to monitor and benchmark their activities, and further enables them to score their performance in key operational areas in a way that leads to enhanced project and corporate performance.

By adopting a holistic approach to customer requirements, building on stakeholder contribution, not only can 'added value' be attained but it can also be measured. Once measured, a benchmark can be set for engaging in a continued drive for organisational excellence.

The MFAM addresses the critical issue addressed by Hersey and Blanchard (1972, cited by Hutchin 2001): 'How do managers cope with the inevitable barrage of changes, which confront them daily in attempting to keep their organisations viable and current? While change is a fact of life, effective managers … can no longer be content to let change occur as it will, they must be able to develop strategies to plan, direct and control change.'

The MFAM provides a means for managers to address the above key question by providing a structured approach to change-management focused upon an organisation's (and its manager's) activities in a drive for corporate excellence.

The incorporation of triple-loop learning enables construction managers to be proactive in relation to change, and this could prove to be most beneficial for all stakeholders.

Developing measurement tools

Given the current attention given to becoming learning construction organisations it seems appropriate that we begin to formalise a measurement method and communications model that will further enhance learning, reinforce positive outcomes and minimise negative outcomes for construction companies.

The model presented is a 'functional assessment model'. However, the functional assessment model forms part of 'competitive-orientated management'. This is a system of management designed to gain and sustain a competitive corporate advantage. The concept of competitive-orientated management may be represented as a tetrahedron, as depicted in *Figure 6.2*. It is based upon three principles of competitive achievement, leadership, personnel and development. Hardy (1983) states that the development of a competitive advantage automatically creates an opportunity, and so the reasoning may be modified to argue that successful businesses are engaged in the creation and exploitation of competitive advantages.

Constituent parts of the tetrahedron

According to Day and Wensley (1988) the essence of competitive advantage is the conversion of superior skills and resources into positional advantages, which in turn create positive outcomes. A competitive advantage is sustained only if it continues to exist after efforts to duplicate that advantage have ceased. Chileshe and Watson (1997, 2000) and Watson and Chileshe (1998) explored the links between TQM and competitive advantage and found that organisations implementing TQM had improved their efficiency and effectiveness. Other sources of competitive advantage may be obtained through the use of benchmarking best practice (Shakantu and Talukhaba 2002), organisational learning, or organisations forming strategic alliances

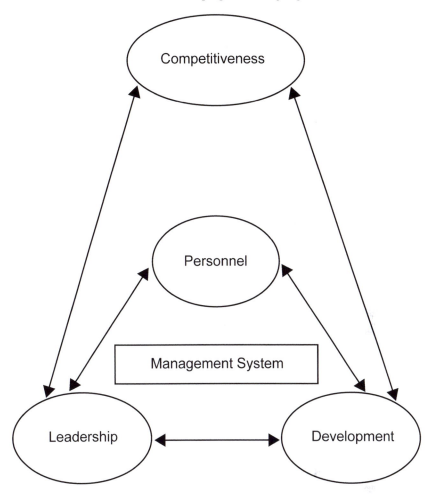

Figure 6.2 Competitive-orientated management core concept

(Ngowi 2001). The following subsection now examines and discusses the constituent parts of the tetrahedron.

Leadership

The role of senior management is critical to the success of any change process and effective leadership has to be demonstrated during the change process. All employees have to be given the time and skills to make a useful contribution towards a drive for a sustainable competitive advantage. 'One factor that affects the individual's reaction to change is their past experiences. Individuals also resist having solutions imposed on them, it is therefore necessary [for construction managers] to obtain the commitment of the individuals to the change initiative' (Kotter 1999, cited by Martin and Cullen 2005).

Personnel

Employees have to be motivated to engage in a corporate culture directed at achieving stakeholder satisfaction. Senior management must not forget that employees are also internal stakeholders of the corporate entity. Thus they should remember to engage in 'consultation' before 'implementation', when engaged in setting and deploying corporate plans. This aspect allows staff to make a valid contribution to the decision-making process. Culture is a vital aspect for consideration and has already been expanded upon in this chapter under the seven functions of management.

Development

A morphogenic culture should be the aim of senior management and the critical issue of staff motivation has to be addressed; this aspect has also been covered under the seven functions of management. The development of a construction learning organisation requires the consideration of both business processes and environmental issues/aspects. Therefore environmental scanning has to be deployed in order to establish relevant external influences, and this may be done by conducting a Strengths, Weaknesses, Opportunities and Threats analysis (SWOT) and/or a Political, Economic, Social, Technological, Environmental and Legal (PESTEL) analysis. These processes should be linked to the RADAR concept; also, full engagement with the concept of a learning organisation is a fundamental requirement for organisational development.

Rationale for MFA Model – development and deployment

Excellence models are strongly related to quality and quality prizes/awards, e.g. the Deming prize in Japan, the Malcolm Baldridge National Quality Award (MBNQA) in the USA and the EFQM Excellence Award in Europe.

Some of these models have been in existence for decades, yet interest in these models has increased. Construction organisations do not just engage with the models in order to win a prize; the models are used in order to guide a firm in achieving organisational excellence. They represent a coherent approach to organisational 'management policies' and help focus a company's attention on critical analytical assessment criteria (Goasdove 2001, cited Hemmel and Ramis-Pujol 2003). As noted by Hemmel and Ramis-Pujol (2003), 'The above models have quickly entered the management practices roller-coaster'.

Various critics have pointed out that some companies who have previously won prizes, have in the long term not performed favourably. For the most part the real reason for these failures is due to poor quality management and inappropriate strategies that are not easily related to by the managers who are responsible for implementing them (Heller 1997, cited Hemmel and Ramis-Pujol 2003).

Beechre and Hamilton (1990, cited Hemmel and Ramis-Pujol 2003) comment that many failures (noting the above) may be attributed to a lack of any attempt at integration and the misalignment of strategic planning, continuous improvement and the transfer of knowledge, when trying to deploy excellence models. The MFAM seeks to address these noted critical issues.

Many construction managers do seem to have great difficulty in understanding concepts, and this could be a contributing factor to existing excellence-model failure. In today's dynamic and very challenging economic operational environments, implementation is viewed not as a choice between options but instead as the ' ... art of balancing among those options'. What to balance are aspects that could be further investigated (Pascale 1992).

The complexity of excellence-model suitability and implementation is further complicated when one adds to it the size of a construction company attempting to deploy such models. The MFA model provides the focus managers require in order to be proactive in the management of change processes. At the same time it also establishes a focal point for linking in a truly holistic manner the sometimes disparate functions of:

- setting and implementing strategic plans;
- setting and implementing operational plans;
- providing due consideration to organisational size, when selecting and engaging in self-assessment linked to an improvement model;
- linking the various functions of management in an effective and efficient way;
- obtaining feedback for stakeholders on organisational performance, with a view to the enhancement of service and product provision;
- building on the concept of triple-loop learning leading to continuous corporate improvement and enhanced customer satisfaction.

Thus the MFA model does address the issues noted above by Hemmel and Ramis-Pujol (2003) and Pascale (1992). The MFAM provides a means of self-assessment for construction organisations related to the seven functions of management. Construction companies can utilise the model in order to establish how efficient and effective they are operating, and further, they can identify areas where they need to improve their project and corporate performance levels.

Management Functional Assessment Model (MFAM) constituent parts

The MFAM is based upon six functions of management: (Forecasting and Planning being linked together under one heading; thus all previously noted seven functions are incorporated). These are Forecasting and Planning, Organising, Motivating, Controlling, Co-ordinating and Communicating.

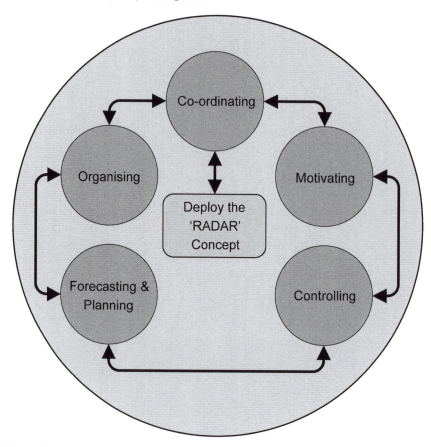

Figure 6.3 Management Functional-Assessment Model (MFAM) incorporating the RADAR concept

The first five functions are encapsulated with a framework of an effective and efficient system of communication (see *Figure 6.3*). The MFAM has been designed to aid managers in determining the key activities to be addressed in order to improve corporate efficiency and effectiveness.

Communicating

Figure 6.3 indicates that communication encompasses the other functions; this is because construction managers engaging in communication are active in a two-way process; also they will be on the receiving end of communications (Fryer 1997).

Forecasting and planning

This criterion is concerned with the shape of future strategy in both the short and the long term; its function is to answer three basic questions:

- where is the company now in terms of its vision and mission;
- where does it want to be as part of its future corporate plans;
- how is the company to achieve its set aims based upon forecasts?

The forecasting and planning criterion contains five basic categories:

1.1 Setting the objectives and strategic planning process in motion.
1.2 Gathering and analysing information related to both clients and the markets (all stakeholders).
1.3 Detailing business processes (who, what, when, where and why).
1.4 Gathering and analysing information relating to competitors and benchmarking.
1.5 Resources planning (incorporating the Five Ms).

Organising

The main managerial tasks here are to organise business processes with a concentration on maximising effectiveness and efficiency. The organising criterion contains five basic categories:

2.1 Creating an appropriate organisational structure.
2.2 Establishing appropriate authority and responsibility for all personnel.
2.3 Creating a self-learning organisational culture (morphogenic/postmodern).
2.4 Developing a corporate value system based upon enhancing performance.
2.5 Deploying new technology linked to corporate enhancement and competitive advantage.

Motivating

It has to be noted that motivation has many aspects both intrinsic and extrinsic, as established when we considered the seven functions of management earlier in this chapter. The motivation criterion contains five basic categories:

3.1 Developing a co-operative culture based on stakeholder satisfaction.
3.2 Ensuring staff have the skills and competencies to perform set tasks.
3.3 A consideration of personnel needs linked to self-actualisation.
3.4 Involvement in processes, increased areas of responsibility and self-monitoring (empowerment).
3.5 Results satisfaction – feedback on performance in a timely manner linked to 'RADAR'.

Controlling

Control is dependent upon constant feedback from each stage of business processes, checking against quality specifications and measuring against performance indicators. A correct monitoring system allows for an increase in the efficiency and effectiveness of organisational activity. Organisations must consider feed-forward of information for effective control and this can only be fully achieved by deploying 'RADAR' which incorporates triple-loop learning. The controlling criterion contains five basic categories:

4.1 A monitoring system for each key stage of business process.
4.2 Measuring performance levels (with an internal and external perspective).
4.3 Determining customer satisfaction levels.
4.4 Determining the efficiency and effectiveness of resource utilisation linked to project and corporate aims.
4.5 Conducting a comparative analysis between set targets and actual results, leading to appropriate and timely actions (RADAR).

Co-ordinating

The analysis of deviations on business processes and updating of the current plans in a holistic manner based on feedback and feed-forward is a critical point in co-ordination. Again this can only be fully attained by the application of 'RADAR'. The co-ordinating criterion contains five basic categories:

5.1 Unity of all the other functions.
5.2 Establishing effective communications.
5.3 Developing a conflict-solving culture, linked to corporate enhancement.
5.4 Updating of deviations: revision and the possible re-co-ordination of resources.
5.5 Information management – information has to be timely and in sufficient detail to inform corrective actions (RADAR).

Communicating

This is the link and the lifeblood of corporate activity and its effectiveness is measured within the context of the other five functions.

In scoring 0–4 the following criteria as shown in Table 6.2 should be applied.

MFAM analysis communication

The presentation of the analysis can be easily communicated to all staff via the application of a communication MFAM pentagonal profile as depicted in *Figure 6.4*. The scores can be plotted upon the profile and a corporate profile established. This process will also demonstrate where a construction organisation should place its efforts in order to improve performance. As an example, *Figure 6.5* contains some fictional data plotted on the profile.

One must remember that action taken in one area will impact upon others. In other words the criteria are not mutually exclusive. Each time the

Table 6.2 Summaries of results for scoring the MFAM

Maturity level	Total score allocated	Assessment results
I	(0– 20)	No methodology or clearly developed processes have been demonstrated. Management's purposes and functions are not clearly defined. For further development it is necessary to reconsider the basic systems and corporate core business principles.
II	(21–40)	A methodology and some processes are in evidence but are erratic in their application. Managers should further develop their leadership skills, define organisational purposes more clearly and develop a strategy based on sound TQM/EFQM EM principles.
III	(41–60)	Management systems and processes are in evidence and utilised, and their approach is evaluated. It is necessary to pay attention to the optimisation of business processes and the improvement of quality at each stage. Need to perfect a control system linked to the importance of stakeholders.
IV	(61–80)	Management systems and processes are in evidence and linked to deployed approaches with constant quality checks within the management system taking place. Utilisation of external benchmarking in order to improve corporate performance has been demonstrated.
V	(81–100)	There is a clear demonstration of a systematic approach to strategy and policy setting linked to the deployment/approach undertaken. The management system is fully functional and the system is benchmarked and monitored in a drive for continuous improvement and the concept of RADAR is fully embedded.

Table 6.3 Scoring criteria to be applied when scoring the MFAM

Score	Criteria for scoring
0	No activity has been demonstrated.
1	Little activity has been demonstrated in this area.
2	Activity utilised but its use is sometimes dependent upon the situation, not a consistent approach.
3	The activity is deployed permanently and systematically.
4	The activity is deployed permanently and systematically, monitored and reviewed via benchmarking for improvement purposes, triple-loop learning is employed via RADAR.

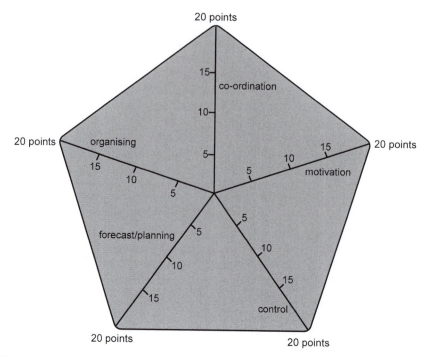

Figure 6.4 Communication MFAM pentagonal profile

MFAM is implemented and corrective actions are taken, a new profile can be developed in order to benchmark the effectiveness of actions taken to improve corporate performance and hence address the competitive advantage issue. This activity becomes part of the 'RADAR' approach.

The advantage of the approach depicted in *Figure 6.5* is in the implementation of benchmarking and feeding forward the results and learning from them. Thus each time the MFA model is deployed it is set within the context of 'RADAR' and hence internal and external benchmarking is inherent. (RADAR has been fully explained in Chapter 4.) From *Figure 6.5* it is clear that this company has issues in relation to Forecasting/Planning and Controling; also Motivating has a low score.

Linking RADAR and the Management Functional Assessment model (MFAM)

The Management Functional Assessment model incorporating RADAR encapsulates the facility for construction organisations to fully engage in a drive for continuous learning and improvement. Every time the MFAM is implemented and the scoring process applied RADAR is embodied within the model. In this way forecasts and plans linked to deployment strategies are evaluated and appropriate actions determined via assessment and review.

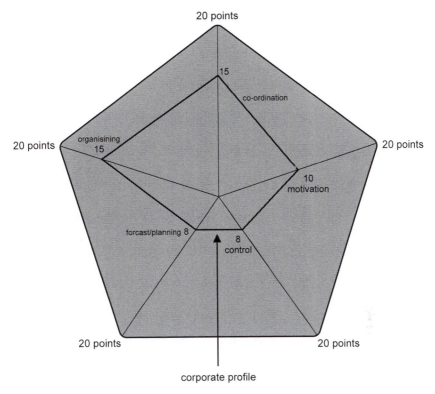

Figure 6.5 Fictitious company's results plotted on the pentagonal profile

Only by employing this approach can the full benefits of MFAM deployment be attained by construction firms.

Figure 6.6 makes clear the critical links between the RADAR concept and the MFAM.

The MFAM recognises that sustainable excellence in all aspects of performance is based on the management functions of Forecasting and Planning, Organising, Co-ordinating, Motivating, Controlling and Communication. The application of the MFAM will address the key development issues shown in *Table 6.4* and empower the resulting benefits.

Excellence is dependent upon balancing and satisfying the needs of all relevant stakeholders and this includes people employed, customers, suppliers and society in general, as well as those within the organisation. The customer is the final judge of product and service quality and customer loyalty, retention and market share are best optimised through a clear focus on the needs of current and potential stakeholders/clients.

Adopting an ethical approach and exceeding the expectations and regulations of the community at large best serve the long-term interests of any construction organisation. Corporate excellence is measured by an organisation's ability to both achieve and sustain outstanding results for its

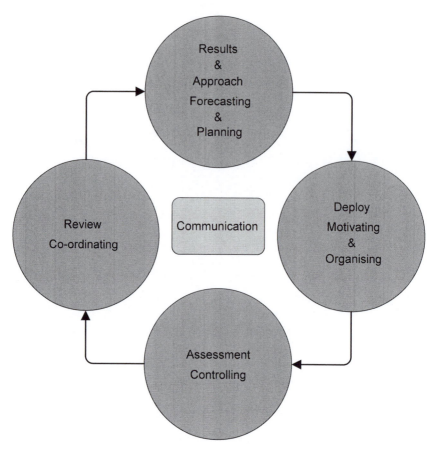

Figure 6.6 MFAM Linked to RADAR

Table 6.4 Deployment of MFAM issues and advantages

Key development issues	Resulting benefits
Process improvements	A clear understanding of how to deliver value to clients and hence gain a sustainable competitive advantage via operations.
Attaining a construction organisation's objectives	Enabling the mission and vision statements to be accomplished by building on the strengths of the company and avoiding any weaknesses.
Benchmarking key performance indicators	Ability to gauge what the construction organisation is achieving in relation to its planned performance targets.
Development of clear, concise action plans resulting in a focused policy and strategy	Clarity and unity of purpose so that the organisation's personnel can excel, learn and continuously improve.

Table 6.4 (continued)

Key development issues	Resulting benefits
Integration of improvement initiatives into normal operational activities	Interrelated activities are systematically managed with a holistic approach to decision making.
Development of group/team dynamics	People development and involvement. Shared values and a culture of trust, thus encouraging empowerment in line with postmodernist/morphogenic company culture.
Embed RADAR	Triple-loop learning and the self-perpetuating culture of learning and continuous project and corporate improvement

stakeholders; thus MFAM linked to RADAR with triple-loop learning has been developed and advocated.

The MFAM has been designed to aid construction managers in determining the key activities to be addressed in order to improve project and corporate efficiency and effectiveness within a framework for obtaining stakeholder satisfaction (Watson, Chileshe and Maslow 2005, Watson, Maslow and Chileshe 2005).

Conclusions

According to Harvey and Denton (1999, cited by Hemmel and Ramis-Pujol 2003), six fundamental developments provide the rationale for the importance and popularity of organisational learning and they are:

- the shifting importance of the factors of production;
- the accelerating pace of change in business environments;
- knowledge is viewed as a source of competitive advantage;
- customers are becoming more demanding;
- dissatisfaction with the existing management paradigm; and
- the increasing intensity of competition.

Further Senge (1999) purported that the real lesson of the quality movement is in the 'learning' that organisations obtain from the process of deployment. In addition, he demonstrated that the evolution of learning organisations can be studied as a series of three quality waves:

- First quality wave, where the primary focus of change was on front-line workers.
- Second quality wave, based on improving management's effectiveness.
- Third quality wave, where learning is institutionalised.

The above points are critical aspects of achieving organisational excellence. The MFAM has been designed to overcome the problematic issues of becoming a 'learning organisation' by the application of 'triple-loop learning'. It has also built in the key components of management, thus overcoming the issues noted above for construction-related companies.

The deployment of the MFA model has to be led (as do all change-management processes) by senior management. Therefore the deployment has to be planned and fully resourced. In fact, it should be subjected to the same monitoring processes as any other change project.

The model (MFA) recognises that sustainable excellence in all aspects of performance is based on the management functions of Forecasting and Planning, Organising, Co-ordinating, Motivating, Controlling and Communication. Excellence is dependent upon balancing and satisfying the needs of all relevant stakeholders (this includes people employed, customers, suppliers, and society in general, as well as those with financial interests in the construction organisation).

Construction firms perform more effectively and efficiently when all interrelated activities are understood and systematically managed and decisions concerning current operations and planned improvements are made using reliable information that includes stakeholder perceptions. So the application of RADAR is essential if a truly holistic control mechanism is to be attained.

Corporate excellence is measured by an organisation's ability to both achieve and sustain outstanding results for its stakeholders; thus MFAM linked to RADAR has been developed. The MFAM model has thus far found favour with the European Foundation for Quality Management and SixSigma – both have published the model.

Case study exercise for the reader: deploying the MFA model

A managing director of a construction company has decided to fully engage with the concept of a learning organisation; however, there exists some resistance to this approach within the host company. The managing director feels that the deployment of the MFAM would be a good starting point for the company. You have been appointed as external consultant and asked to prepare a presentation to convince the rest of the board of directors (based upon this chapter). Your presentation should consist of key bullet points, and these should relate to the advantages of deploying the MFAM.

When complete you can compare your list with that provided at the end of the book.

Question 4 for the reader

The Management Functional Assessment Model (MFAM) provides a focal point for managers seeking to be proactive in the management of change processes. The MFAM also combines six functions that are usually considered in a disparate fashion, if at all. Identify these six functions. An answer is provided at the end of the book.

Further reading

Goh, S. C., (1998). 'Towards a Learning Organisation: The Strategic Building Blocks'. SAM. Advanced Management Journal, Vol. 63, No. 2, pp. 15–20.

Hermel, P. H., (1997). 'The New Focus of Total Quality in Europe and the US'. *Total Quality Management*. Vol. 8. No. 4, pp. 131–43.

McCabe, S., (1998). *Quality Improvement Techniques in Construction*, Longman, Edinburgh Gate.

Oakland, J., Tanner, S. and Gadd, K., (2002). Best practice in business excellence. *Total Quality Management*, 13(8), pp. 1125–39.

Patton, R, A., and Mccalman, J., (2000). *Change Management: A Guide to Effective Implementation*. 2nd ed. London: Sage.

Watson, P., (2002). Developing an Efficient and Effective Control System. *Journal of the Association of Building Engineers*. Vol 77. February.

References

Argyris, C. and Schon, D. A., (1974). *Theory in Practice: Increasing Professional Effectiveness*. San Francisco: Jossey-Bass.

——(1978). *Organisational Learning: A Theory of Action Perspective*. Reading, MA: Addison – Wersley.

Barlow, J., and Jashspsra, A., (1998). *Organisational Learning and Inter-firm 'Partnering' in the UK Construction Industry* [online]. The Learning Organisation, 5(2), pp. 86–98. Available from: Emerald Fulltext Database [Accessed 9 November 2004].

Bessant, J (1998), 'Learning and continuous improvement', in Tidd, J. (ed.), *Measuring Strategic Competencies: Technological, Market and Organizational Indicatiors of Innovation*, Imperial College Press, London.

Bessant, J, Caffyn, S (1997), 'High involvement innovation', *International Journal of Technology Management*, Vol. 14, No.1, pp. 7–28.

Bounds, G., Yorks, L., Adams, D. M. and Ranney, G., (1994). *Beyond Total Quality Management: Towards the Emerging Paradigm*. McGraw-Hill International Editions.

Burns, T. and Stalker, G. M., (1961). *Managing Innovation*. London: Pergammon.

Cole, G., (1995). *Organisational Behaviour*. Continuum.

Cummings, A., (2005). *A Critical Examination of the Key Issues that Influence a Large Construction Company's Ability to Become a Learning Organisation*. BSc (Hons) Dissertation, Sheffield Hallam University.

Dahlgaord, S. M. P., (2004). Perspectives on Learning in a Literature Review. *European Quality*. 2(1), pp. 033–47.

De Genus, A.P., (1988). Planning as Learning. *Harvard Business Review*, 66(2), pp. 70–74.

European Foundation for Quality Management (1999). *Radar and the EFQM Excellence Model*, EFQM Press Releases & Announcements, [online]. Available at: http://www.efgm.org [accessed 12 June 2000].

Fryer, B., (1997). *The Practice of Construction Management*. 3rd ed. Blackwell Science, Ltd.

Garvin, D. A., (1993). Building a Learning Organisation. *Harvard Business Review.* 71(4), pp.78–91 [online]. Available from: EBSCO Business Source Premier Database [accessed 4 November 2004].

Greising, D., (1994). Quality: How to Make It Pay. *Business Week*, (8 August), pp. 54–59.

Griffith, A. and Watson, P., (2004). *Construction Management: Principles and Practice.* Basingstoke: Palgrove Macmillan.

Hamel, G. and Prahalad, C. K., (1994). *Competing for the Future.* Boston: Harvard Business School Press.

Hardy, L., (1983). *Successful Business Strategy – How to win the market place.* Kogan Page.

Hemmel, P. and Ramis-Pujol, J., (2003). An Evolution of Excellence: Some Main Trends. *TQM Magazine*, 15(4), pp. 230–43.

Ho, S. K. M., (1999). *Total Learning Organisation.* The Learning Organisation, 6(3), pp. 116–20 [online]. Available from: Emerald Fulltext Database [accessed 4 November 2004].

Hutchin, T., (2001). *Unconstrained Organisations: Managing Sustainable Change.* Thomas Telford Ltd.

Imai, K (1987), Kaizen, Random House, New York, NY.

Jackson, N. and Carpenter, P., (2000). *Rethinking Organisational Behaviour.* Harlow: Prentice Hall, Pearson Education Ltd.

Johnson, G. and Scholes, K., (2002). *Exploring Corporate Strategy:* Text and Cases. 6th ed. Harlow: Pearson Education Ltd.

Leonard-Barton, D. (1992), 'The organisation as learning laboratory', *Sloan Management Review*, Vol. 34, No.1, pp. 23–38.

Lobermans, J., (2002). Synergising the Learning Organisation and Knowledge Management. *Journal of Knowledge Management*, 6, pp. 285–94 [online]. Available from: Emerald Fulltext Database [accessed 5 November 2004].

Love, P. E. D., (2004). Nurturing a Learning Organisation in Construction: a Focus on Strategic Shift, Organizational Transformation, Customer Orientation and Quality Centred Learning. *Construction Innovation*, 4(2), pp.113–26 [online]. Available from: EBSCO Business Source Premier Database [accessed 16 November 2004].

Majaro, S., (1992). *Managing Ideas for Profit.* Maidenhead: McGraw-Hill Book Company.

Malhotra, Y., (1996). *Organisational Learning and Learning Organisations: An overview.* [Online]. Available at: Http://www.kmbook.com/orglrng.htm [accessed 5 November 2004].

Martin, J. and Cullen, P., (2005).*When precedents are not enough: creating consistent contracts in a changing organisation.* Proceedings of the International Commercial Management Symposium, 7 April. The University of Management, UK.

McCabe, S., (2001). *Benchmarking in Construction.* London: Blackwell Science.

Morris, D. and Brandon, J. (1993). *Re-Engineering Your Business.* London: McGraw-Hill.

Naoum, S., (2001). *People and Organisational Management in Construction.* London: Thomas Telford Ltd.

Newell, S., (2004). Enhancing Cross-Project Learning. *Engineering Management Journal*, 16(1), pp. 12–20. [Online]. Available from: EBSCO Business Source Premier Database [accessed 16 November 2004].

Ngowi, A. B., (2001). The competition aspect of construction alliances. *Logistics Information Management*, 14(4), 242–49.

Nonaka, I., (1991). The Knowledge – Creating Company. *Harvard Business Review*, 69(6), pp. 96–104. [Online]. Available from: ESBCO Business Source Review Database [accessed 7 November 2004].

Nyhan, B. (2004). European Perspectives on the Learning Organisation. *Journal of European Industrial Training*, 28(1), pp. 67–92 [online]. Available from: Emerald Fulltext Database [accessed 16 November 2004].

Ortenbland, A., (2004). *The Learning Organisation: Towards an Integrated Model*. The Learning Organisation 11, pp.129–44. [Online]. Available from: Emerald Fulltext Databse [accessed 4 November 2004].

Pascale, R. T., (1992). *Les risques de l'excellence. La Strategie de conflicts constructifs*. Paris: Inter Editions.

Passmore, W. A., (1994). *Creating Strategic Change*. J Wiley & Sons Inc.

Pedlar, M. and Aspinwall K. *A Concise Guide to the Learning Organisation*. London: Lemos and Crane, 1998.

Pedler, M., Boydell, T., and Burgoyne, J., (1988). *Learning Company Project Report*. Sheffield: Manpower Services Commission.

Peters, T., (1988a). *Thriving on Chaos*. London: Macmillan.

——, (1988b). 'Facing up to the need for a management revolution'. *California Management Review*. Winter, pp. 7–38.

Revans, R. W., (1983). *ABC of Action Learning*. Bromley: Chartwell Bratt.

Robinson, A (1991), *Continuous Improvement In Operations*, Productivity Press, Cambridge, MA.

Sayles, L. R., (1989). *Leadership Managing in Real Organisations*. London: McGraw-Hill, London.

Senge, P. M., (1990). *The Fifth Discipline: The Art and Practice of the Learning Organisation*. London: Century Business.

——(1999). *Association for Quality & Participation*. November/December, pp. 34. 40. 'It's the Learning: The Real Lesson of Quality Movement'.

Shakantu, W. and Talukhaba, A., (2002). Benchmarking Best Practice to Achieve a Competitive Advantage in the South African Construction Industry. In: Ahmed, S. M., Ahmad, I., Tang, S. I. and Azhar, S., eds, *First International Conference on Construction in the 21st Century 'Challenges and Opportunities in Management and Technology'* (CITC-I), 25–26 April 2002, Florida International University, Miami, USA.

Schroeder, M, Robinson, A. (1993), 'Training, continuous improvement and human relations: The US TWI programs and Japanese management style', *California Management Review*, Vol. 35, No.2.

Shukla, M., (1997). *Competing Through Knowledge: Building a Learning Organisation*. New Delhi: Response Books.

Stata, R., (1989). Organisational Learning – The Key to Management Innovation. *Sloan Management Review*. Spring.

Sun, P. Y. T. and Scott, J. L., (2003). Exploring the Divide: Organisational Learning and Learning Organisation. *The Learning Organisation*, 10(4), pp. 2002–2215 [online]. Available from: Emerald Fulltext Database [accessed 4 November 2004].

Teece, D and Pisano, G. (1994), 'The dynamic capabilities of firms: an introduction', *Industrial And Corporate Change*, Vol. 3, No. 3, pp. 537–55.

Tidd, J., Bessant, J. and Pavitt, K. (1997), *Managing Innovation: Integrating Technological, Organizational And Market Change*, John Wiley, Chichester.

Watson, P. and Chileshe, N., (1998). *Aspects of Total Quality Management (TQM) Implementation within a Construction Operational Environment*. South African First Congress on Total Quality Management in Construction, November pp. 101–14.

Watson, P., Chileshe, N. and Maslow, D., (2005). *Addressing Sustainable Competitive Advantage via a Functional Assessment Model*. International Commercial Management Symposium. 7 April. University of Manchester, UK.

Watson, P., Maslow, D. and Chileshe, N., (2005). *Management Assessment for Competitive Advantage*, iSixSigma Insights, 7 March 2005, 5(19).

Weber, M., 1908 (1968). *Economy and Society*. Translated and Edited by Roth, G. and Witrich, C., New York: Irving Publications.

7 Quality management systems for health and safety in construction

Introduction

Excellence in health and safety management is an essential attribute of successful modern-day construction organisations. Poor health and safety management can impact upon a construction organisation's reputation, the timely progress of its projects, the morale and commitment of its workforce and the size and future surety of its order book.

This chapter serves to inform of occupational health and safety management systems and outlines the essential components of such systems for organisations. Advocated benefits and problems associated with occupational health and safety management systems are indicated and differing standards and guidance documents are introduced. The application of a systematic health and safety management approach to construction projects through compliance with the Construction (Design and Management) Regulations 2007 is highlighted. Examples of useful documentation for contributing to the systematic management of health and safety on construction projects are provided at the end of the chapter.

Learning outcomes

Upon completion of this chapter the reader will be able to demonstrate an understanding of:

- Essential components of occupational health and safety management systems (OHSMS).
- Advocated benefits and problems associated with occupational health and safety management systems.
- An awareness of different health and safety management standards and guidance documents.
- Issues associated with developing an OHSMS within an organisation.
- How compliance with the Construction Design and Management Regulation 2007 facilitates a health and safety management systems approach to the construction.

Essential components of occupational health and safety management systems

Like many management models, occupational health and safety management systems (OHSMS) are commonly founded upon Deming's dynamic control loop cycle. This cycle is illustrated in *Figure 7.1*.

The Institution of Occupational Safety and Health (IOSH) (2009) express the key components of occupational health and safety management systems in terms of a Deming 'Plan-Do-Check-Act' diagram, as illustrated in *Figure 7.2*.

The Institution of Occupational Safety and Health (IOSH 2009) also suggest that effective health and safety management systems contain the following elements:

- Policy – The organisation's statement of commitment and vision. Senior management must lead this policy and be accountable for it.
- Planning – This should address how legal requirements are identified, how hazards are identified and the resultant risks assessed and controlled. It should also document preparation regarding planning for and responding to emergencies.
- Organising – The organisation's structure needs to be defined and health and safety clearly allocated in a manner linked to operational controls. Furthermore there need to be ways of delivering and ensuring awareness, competence, consultation and training.
- Workers /employee representatives – Such representation can invaluably facilitate the health and safety management of the organisation, particularly with regard to improvement opportunities and risk management.
- Communicating – It is essential that this is two-way between the organisation's managers and workers, is regular and ongoing and includes health and safety information relating to work procedures and all aspects of the organisation's OHSMS.
- Consulting – all stakeholders of the organisation need to be identified and consulted effectively regarding health and safety in order to proactively access their knowledge, views, requirements and expertise as well as the reactive feedback.
- Implementing and operating – the OHSMS needs to be put into practice, in its entirety.
- Measuring performance – This can be undertaken by evaluating data relating to incidents (accidents and near misses) and ill health, as well as information obtained from, amongst other things, hazard identification, risk assessments, regular inspections, health and safety committees and training activities.
- Corrective and preventive actions – there must be a systematic approach to proactively preventing incidents, accidents and ill health, as well as corrective measures implemented from the investigation of incidents, accidents and ill health.

- Management review – this must be done in order to ensure compliance with legal requirements and appraise the performance achieved against objectives set, and to re-evaluate the system itself and its resourcing.
- Continual improvement – a commitment to proactively manage health and safety risks is at the core of the system in order to effectively reduce incidents of ill health and accidents with the efficient deployment of reduced resources.

The United Kingdom's Health and Safety Executive (HSE) documents the essential components of a successful health and safety management system in the publication 'Successful Health and Safety Management'. The HSE outlines the following as being essential components of successful health and safety management:

- A clear policy for health and safety.
- Organisation of all employees for the management of health and safety.
- Planning for health and safety – via setting objectives and targets, identifying hazards, assessing risks and establishing standards for the performance of the organisation to be measured against.
- Measurement of health and safety performance.
- Informed improvement by auditing and reviewing safety performance and practice.

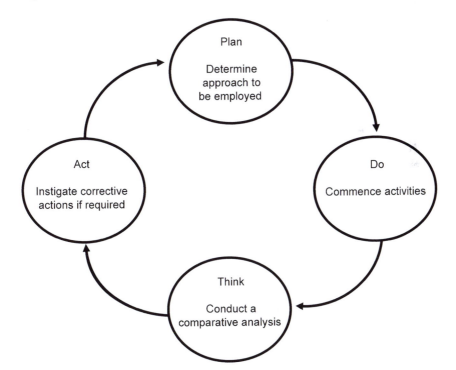

Figure 7.1 Deming's dynamic control loop cycle

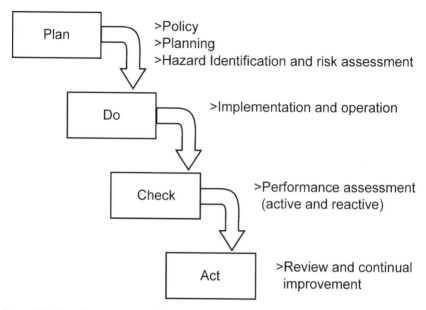

Figure 7.2 Key Components of the OHS management system, according to IOSH

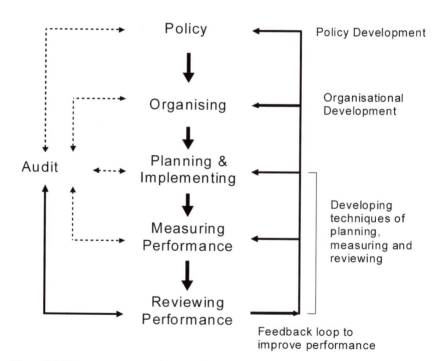

Figure 7.3 Key components of successful health and safety identified by the HSE

The International Labour Office (ILO) has also identified key components of an occupational health and safety management system. The ILO considers that an organisation's system should contain policy, organising, planning and implementation, evaluation and action for improvement. These components are required to be structured in a manner rooted in Deming's dynamic control loop cycle, as illustrated in *Figure 7.4*.

The European Agency for Safety and Health at Work (2002) suggest that an ideal occupational health and safety management system should include a number of key processes. Table 7.1 presents these key processes.

A further occupational health and safety management system is provided by the Occupational Health and Safety Assessment Series (OHSAS) 18001 management systems standard. This standard is commonly recognised around the globe as a leading health and safety management systems

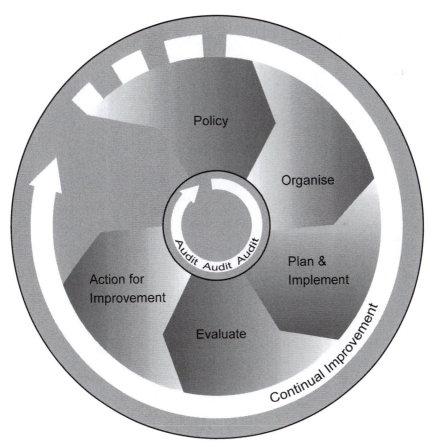

Figure 7.4 Key components of the OSH management system, according to the international labour office

Table 7.1 Key processes of an occupational health and safety management system, according to the European Agency for Safety and Health at Work (2002)

OHS Input – Initiation

20. Management commitment and resources
21. Regulatory compliance and system conformance
22. Accountability, responsibility and authority
23. Employee participation

OHS Process

Implementation	**Formulation**
11. Training system	15. OHS Policy / goals and objectives
12. Hazard control system	16. Performance measures
13. Preventive and corrective action system	17. System planning and development
14. Procurement and contracting	18. Baseline evaluation and hazard assessment
	19. OHSMS manual and procedures

OHS Output Feedback

6. OHS goals and objectives
7. Illness and injury rates
8. Workforce health
9. Changes in efficiency
10. Overall performance of the organisation

OHS Feedback Evaluation

4. Communication system – including document and record management system
5. Evaluation system – auditing and self inspection, incident investigation root-cause analysis, health/ medical programme and surveillance

---

Open Systems Elements

1. Continuous improvement
2. Management review
3. Integration

---

standard. Section 4.1 of OHSAS 18001:2007 sets out five general requirements for an organisation's occupational health and safety management system:

- Establish a management system
- Document the management system
- Implement the management system
- Maintain the management system
- Continually improve the management system

Benefits and problems associated with OHS management systems

Quantifying the benefits and problems associated with the establishment and delivery of an organisation's OHSMS can be challenging. A key qualitative benefit that can be readily associated with the implementation of an OHSMS, though, is the very visible, strategic and operational *commitment* provided to the health, safety and well-being of the organisation's workforce community. An OHSMS can also assist in the delivery of a number of the organisation's legal and moral obligations and it can also facilitate an internal focus on good management practice and continuous improvement.

A number of benefits and problematic issues can be associated with the deployment of OHS management systems. Benefits can include:

- Improved prevention of occupational injury and disease – a safer and healthier workplace.
- The provision of a framework for identifying hazards and managing the resultant risks.
- A reduction in the loss of working days due to accidents and injury.
- A reduction in the incidences of employee compensation claims.
- The development of a reviewable approach for meeting legislative requirements, duties of care and due diligence.
- A reduction in insurance premiums.
- Improvements in morale and productivity brought about by employee inclusivity with developing and running.
- Enhanced working methods that facilitate improvement in production and productivity rates.
- Enhanced reputation of the organisation with a visible and tangible commitment to continuous improvement and inclusive, consultative management mechanisms.
- Reduced staff turnover and thereby reduced 'replacement costs'.
- Improved ability to attract skilled personnel.
- Improved commercial potential – inclusion on tender lists is increased as the potential for meeting the pre-qualification requirements of significant clients is enhanced.

These benefits can be paralleled with those associated with the implementation of a workplace health promotion programme within an organisation. *Figure 7.5* illustrates the 'framework for describing arguments based on the effects and outcomes of workplace health promotion' as prescribed by the European Agency for Safety and Health at Work (2009).

Problems that can be associated with the development and delivery of an OHSMS include:

- The system is not organisation specific. The system needs to be tailored to the organisation and its culture. There is no ready-made 'off the shelf' solution.
- Management support may be lacking. The leadership and commitment of management needs to be visible and suitable management priority needs to be given to establishing, developing and improving the system.

- Understanding of the purpose and benefits of the system is lacking – here the system can be viewed as a 'paper trail' or hindrance to daily work, potentially with procedures put in place in a top-down manner.
- The OHSMS is established due purely to external drivers – possibly in order to enable inclusion on client tender lists. With this external driver alone the system will rarely achieve the necessary 'ownership' by those internal to the organisation.
- There is insufficient 'ownership' of the system from persons across the organisation. Without broad and effective participation in development, sustained delivery and improvement, the system will be viewed as one that is 'imposed'.

Health and safety management standards and guidance documents

Common to the various occupational health and safety management standards and guidance documents that have been developed is that each is based upon Deming's dynamic control loop cycle. Furthermore each shares the intent of facilitating the delivery of robust, systemised occupational health and safety management practice.

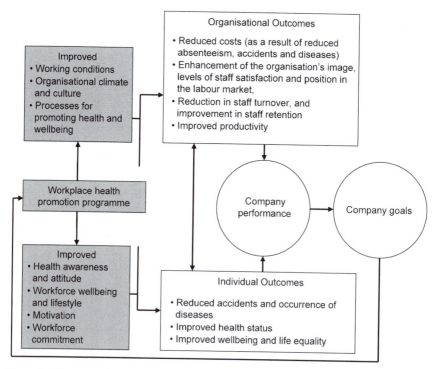

Figure 7.5 Framework for describing arguments based on the effects and outcomes of workplace health promotion

When undertaking to develop and implement an occupational health and safety management system within an organisation a number of 'standards' are worthy of consideration and consultation; these include:

- HSG65 Successful Health and Safety Management
- ILO OSH: 2001 (Guidelines on Occupational Health and Safety Management Systems)
- OHSAS 18001:2007 Occupational health and safety management systems – requirements
- OHSAS 18002:2008 Occupational health and safety management systems – Guidelines for the implementation of OHSAS 18001:2007
- BS 18004:2008 Guide to achieving effective occupational health and safety performance

Knowledge of each of these occupational health and safety standards is worthwhile, especially when undertaking to develop and implement an occupational health and safety management system. The standards do not prescribe what an organisation must do; instead the standards provide a 'framework' to help key issues to be identified in order that the system that is developed is suitably and effectively aligned with the host organisation.

Figure 7.6 presents a timeline overview of the development of various key occupational health and safety management standards.

HSG65 successful health and safety management

'Successful health and safety management' was initially prepared by the Health and Safety Executive's Accident Prevention Advisory Unit to provide guidance for directors, managers and safety professionals who were seeking to improve health and safety performance. It was first published in 1991 and has since been revised. It is not an approved code of practice or a certifiable international standard. It is a guidance document.

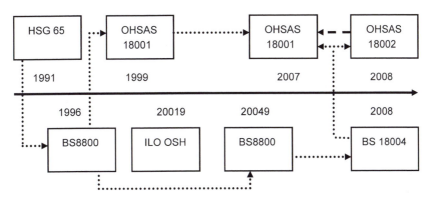

Figure 7.6 Timeline of occupational health and safety management 'standards'

In presenting sound guidance on the practice of health and safety management within organisations, the following content is addressed:

- Effective health and safety policies
- Organising for health and safety
- Planning and implementing
- Measuring performance
- Auditing and reviewing performance

The POPMAR model for managing health and safety is a well recognised feature of this systematic guidance. Here Policy, Organisation, Planning and implementing, Measuring, Auditing and Reviewing are presented as the key elements of a successful health and safety management system.

ILO OSH: 2001 guidelines on occupational health and safety management systems

The International Labour Organisation (ILO) issued ILO-OHS 'Guidelines on Occupational Health and Safety Management System' in 2001, further to consultation and a meeting of experts in April 2001. The guidelines are intended for application at two levels, a national level and an organisational level. The organisational level seeks to:

- Provide guidance regarding the integration of OHS management system elements in the organisation as a component of policy and management arrangements; and
- Motivate all members of the organisation, particularly employers, owners, managerial staff, workers and their representatives, in applying appropriate OHS management principles and methods to continually improve OHS performance (ILO OHS 2001).

The guidelines are not legally binding and their application does not necessitate certification. The following content is addressed by the guidelines:

- Policy
- Organising
- Planning and implementation
- Evaluation
- Action for Improvement

OHSAS 18001:2007 Occupational health and safety management systems – requirements

OHSAS 18001 documents requirements for an occupational health and safety management system and serves to support organisations to control OH&S risks and improve performance. OHSAS 18001:2007 replaces OHSAS 18001:1999.

The standard has been adopted by British Standards. BS OHSAS 18001 states the requirements that must be met to demonstrate that an organisation has an effective occupational health and safety management system. It is a standard for which certification for compliance can be sought. This enables an organisation to demonstrate via independent audit to their stakeholders that the organisation operates a health and safety management systems with elements and characteristics that are in accordance with the standard.

The standard is supported by OHSAS BS 18002:2008 Guidelines for the implementation of OHSAS 18001:2007 and BS 18004:2008 Guide to achieving effective occupational health and safety performance.

Baker (2001) argues that whilst organisations can see the value of an OHSMS such as OHSAS 18001, there can be perceived limitations associated with such a standard. The certification of OHSAS 18001, it can be argued, could indicate a good level of safety administration rather than effective safety and health management – it can be regarded as focusing on written documentation.

Table 7.2 provides an overview of the content of this safety standard.

OHSAS 18002:2008 Occupational health and safety management systems – guidelines for the implementation of OHSAS 18001:2007

These guidelines serve to support the implementation of OHSAS 18001 and explain the principles of OHSAS 18001. This standard serves to assist organisations understand and implement OHSAS 18001 through the provision of examples and aspects to consider when undertaking to implement or audit OHSAS 18001.

BS 18004:2008 Guide to achieving effective occupational health and safety performance

This standard replaces BS8800 2004 and serves to add to the requirements of 18001 and the guidance in 18002 by providing more detailed information about key elements of effective occupational health and safety management. BS 18004 is for organisations who seek to:

- Establish an OH&S management system to control risks to personnel and other interested parties who could be exposed to OH&S hazards associated with its activities.
- Implement, maintain and continually improve the OH&S management system.
- Demonstrate commitment to good practice, including self-regulation and continuous improvement in OH&S performance.
- Assure conformity with the H&S policy and BS OHSAS 18801 by either self-determination and a declaration, seeking confirmation from either an organisational stakeholder or an external party, or via certification of the OHSMS by an external organisation.

Table 7.2 Overview of OHSAS 18001:2007

1	Scope
2	Reference publications
3	Terms and definitions
3.1	Acceptable risk
3.2	Audit
3.3	Continual improvement
3.4	Corrective action
3.5	Document
3.6	Hazard
3.7	Hazard identification
3.8	Ill health
3.9	Incident
3.10	Interested party
3.11	Nonconformity
3.12	Occupational health and safety (OH&S)
3.13	OH&S management system
3.14	OH&S objective
3.15	OH&S performance
3.16	OH&S policy
3.17	Organisation
3.18	Preventive action
3.21	Risk
3.22	Risk assessment
3.23	Workplace
4.1	General requirements
4.2	OH&S policy
4.3.1	Hazard identification, risk assessment and determining controls
4.3.2	Legal and other requirements
4.3.3	Objectives and programme(s)
4.4.1	Resources, roles, responsibility, accountability and authority
4.4.2	Competence, training and awareness
4.4.3.1	Communication
4.4.3.2	Participation and consultation
4.4.4	Documentation
4.4.5	Control of documents
4.4.6	Operational control
4.4.7	Emergency preparedness and response
4.5	Checking
4.5.1	Performance measurement and monitoring
4.5.2	Evaluation of compliance
4.5.3	Incident investigation, nonconformity, corrective action and preventive action
4.5.3.1	Incident investigation
4.5.3.2	Nonconformity, corrective and preventive action
4.5.4	Control of records
4.5.5	Internal audit
4.6	Management review

Furthermore the standard also provides guidance on promoting an effective OH&S management system and investigating hazardous events.

Developing an occupational health and safety management system

When undertaking to develop and implement an OHSMS such as OHSAS 18001:2007, it is necessary for the host organisation to rigorously align the H&S standard with the organisation. This tailored alignment necessitates considered documentation, so as to provide evidence of the critical process of the constructive application of the standard to the host organisation. This process requires the identification of H&S aspects or 'inputs to the system'. The Institution of Occupational Safety and Health (IOSH) outline a process for the development within an organisation of an occupational health and safety management system. This process is arranged into three sections, these being 'typical inputs', a 'gap analysis review' and the 'development of a draft management system' that documents a number of key components. IOSH's process for developing an OHSMS is presented in *Figure 7.7*.

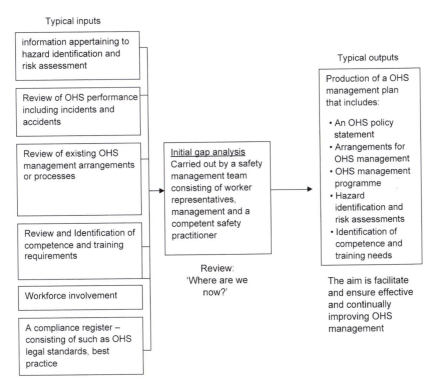

Figure 7.7 Process for developing an OHSMS
(IOSH 2009, p. 23)

Developing documentation – a case study example

When undertaking to develop and document a management system a 'Consensus Chart' is proposed by Laman (2009) in order to classify issues and improve the process. This chart is presented in *Figure 7.9* and is completed with safety and health issues of a one-example organisation completed within the chart matrix. Application of this consensus chart approach is considered to facilitate a standardised and effective process for obtaining buy-in and comprehensive documentation. Laman considers the chart to be especially useful in an environment, such as that of an organisation's OHSMS, with the following characteristics:

- Continuous improvement
- Differences of opinion
- Desire to optimise documentation
- Appreciation of the team approach
- Complex processes that interact with other processes

In applying the 'consensus chart' approach to the documentation development process a representation of management and employees is necessary. This representation is required to give due consideration to identifying health and safety issues and the current context of each issue within the organisation.

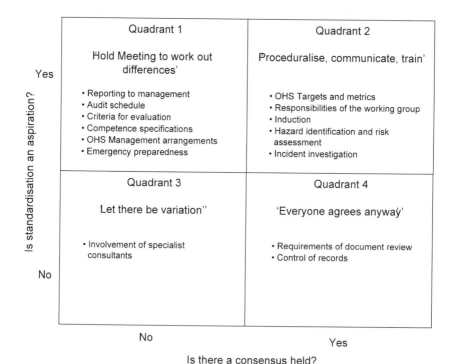

Figure 7.8 Laman's consensus chart

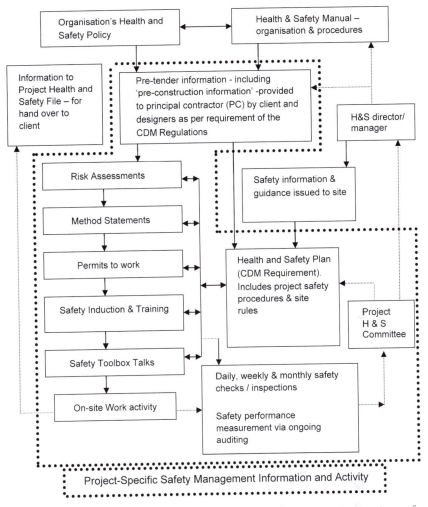

Figure 7.9 Key components of a principal contractor's construction project safety management system

Laman suggests posing two questions during this process:

- Do we want to standardise?
- Is there consensus about best practices?

The answers, or 'issue responses', provided by the representative group, are located within one of the four 'action' quadrants of the consensus chart.

- 'Hold meetings to work out differences' – Here standardisation is desired, but consensus has not yet been achieved.
- 'Proceduralise, communicate, train' – There is both desire and consensus to standardise. Process improvements can be initiated.

- 'Let there be variation' – There is no desire to standardise and consensus is not held. No action results from this outcome.
- 'Everyone agrees anyway' – Consensus exists but there is no need to standardise.

Where standardisation is agreed – 'Proceduralise, communicate, train' is the outcome – new documentation will result. This documentation could include new and revised policies, procedures, forms and training plans.

Measuring performance of an organisation's occupational health and safety management system

The development and implementation of an OHSMS can bring about reductions in incidents, accidents and ill health. Furthermore it can reduce resultant costs, lost time and insurance premiums as well as improve reputation, well-being and motivation. This is all well and good but can lead to organisations focusing attention upon metrics that narrowly concern safety performance resulting or trailing from the OHSMS. It is imperative that the performance of the system itself is measured and monitored in terms of metrics that seek to deliver efficiency and continuous improvement.

Warren (2005) – in European Agency for Safety and Health at Work, 2009 – suggests that the performance measurement of management systems must be 'SMART': Specific, Measurable, Achievable, Relevant and Time-based.

- *Specific*:
  Performance criteria should be as specific as possible to make sure that it is easy to identify what is being measured.
- *Measurable*:
  Performance criteria need to be measurable, either in quantity or by quality, to check that stipulated goals are being met.
- *Achievable*:
  Unrealistic goals may cause disease within an organisation. However, the challenge of goals that stretch an organisation a little may be beneficial.
- *Relevant*:
  The performance measurements should be relevant to the organisation's overall mission and to the strategic objectives of any programme.
- *Time-based*:
  The performance measurements should be achievable within a specific period.

These five performance measurement characteristics provide helpful direction for the development of performance metrics for a safety management system.

Street (2000) suggests that performance metrics can be categorised into three types: 'trailing indicators', 'current indicators' and 'leading indicators', as illustrated in *Table 7.3*. Together these three categories of metric provide a

Table 7.3 The CDM2007 duties of parties to a construction project

	Duties for all construction projects (Part 2 of the regulations)	Additional duties for notifiable projects (Part 3 of the regulations)
Clients	• Check competence and resources of all appointees • Ensure there are suitable management arrangements for the project including welfare facilities • Allow sufficient time and resources for all stages • Provide pre-construction information to designers and contractors	• Appoint co-ordinator* • Appoint principal contractor* • Make sure that the construction phase does not start unless there are suitable: • welfare facilities, and • construction phase plan • Retain and provide access to the health and safety file (*There must be a co-ordinator and principal contractor until the end of the construction phase)
CDM Co-ordinators		• Advise and assist the client with his/her duties • Notify HSE • Co-ordinate health and safety aspects of design work and cooperate with others involved with the project • Facilitate good communication between client, designers and contractors • Liaise with principal contractor regarding ongoing design • Identify, collect and pass on pre-construction information • Prepare/update health and safety file
Designers	• Eliminate hazards and reduce risks due to design • Provide information about remaining risks	• Check client is aware of duties and CDM co-ordinator has been appointed • Provide any information needed for the health and safety file

Table 7.3 (continued)

	Duties for all construction projects (Part 2 of the regulations)	Additional duties for notifiable projects (Part 3 of the regulations)
Principal contractors	N/A	• Plan, manage and monitor construction phase in liaison with contractors; • Prepare, develop and implement a written plan and site rules. (Initial plan completed before the construction phase begins.) • Give contractors relevant parts of the plan • Make sure suitable welfare facilities are provided from the start and maintained throughout the construction phase • Check competence of all their appointees • Ensure all workers have site inductions and any further information and training needed for the work • Consult with the workers • Liaise with co-ordinator re ongoing design • Secure the site • Check client is aware of duties and a co-ordinator has been appointed and HSE notified before starting work • Co-operate with principal contractor in planning and managing work, including reasonable directions and site rules • Provide details to the principal contractor of any contractor who he engages in connection with carrying out work • Provide any information needed for the health and safety file • Inform principal contractor of problems with the plan • Inform principal contractor of reportable accidents, diseases and dangerous occurrences
Contractors	• Plan, manage and monitor own work and that of workers • Check competence of all their appointees and workers • Train own employees • Provide information to their workers • Comply with the specific requirements in Part 4 of the regulations • Ensure there are adequate welfare facilities for their workers	

Table 7.3 (continued)

	Duties for all construction projects (Part 2 of the regulations)	Additional duties for notifiable projects (Part 3 of the regulations)
Everyone	Check own competence • Co-operate with others and co-ordinate work so as to ensure the health and safety of construction workers and others who may be affected by the work • Report obvious risks • Comply with requirements in Schedule 3 and Part 4 of the regulations for any work under their control • Take account of and apply the general principles of prevention when carrying out duties	

(Table extracted from HSE 2007, Managing health and safety in construction. Construction (Design and Management) Regulations 2007. Approved Code of Practice. Pg 6. HSE Books)

comprehensive platform of data and information from which a safety management system can be viewed and evaluated.

When measuring the performance of an OHSMS it is important to appreciate the distinction between 'incidents' and 'nonconformities'. Section 4.5.3 of OHSAS 18001:2007 concerns 'incident investigation, nonconformity, corrective action and preventative action.' OHSS 18001 defines an incident as a 'work-related event in which an injury or ill health or fatality occurred, or could have occurred'. An 'accident' is a particular category of incident, one in which injury or illness results, regardless of severity. Another category of incident is one where no illness or injury results; this is termed a 'near miss'.

It is possible for an organisation to be operating in a manner of nonconformity with its OHSMS. Here aspects such as work procedures, information flows, worker engagement and consultation, training, risk assessment or the like may not be in full accordance with the developed OHSMS. An organisation may have nonconformities without the immediate occurrence of incidents. Such nonconformities indicate insufficient attention to implementation, control, review and ownership of the system.

Auditing the system

Audit is a necessary component of any system that seeks to ensure conformity and continuous improvement. The performance metrics utilised for the evaluation of the system can provide a worthwhile data set for the monitoring and managed improvement of the organisation's safety management system. The Health and Safety Executive prescribe a worthwhile checklist tool for the auditing of eight components of an organisation's health and safety management system (HSE 2010). The tool enables an organisation to assess their health and safety management system by 'self-scoring' against specified elements. By repeating the self-assessment exercise after improvements have been made to the system, progress can be measurably scored and recorded over time. *Table 7.4* presents this self-assessment audit tool.

The health and safety management system and the construction project

Building and infrastructure project workplaces are dynamic and diverse in many respects. They are procured in a variety of ways, have various contractual arrangements, differing project management organisational structures and present many technical innovations and challenges. It is imperative that a project health and safety management system can fully address the various challenging arrangements and aspects of differing projects in a systematic manner. The principles of approach for a project safety management system are the very same as for the development and implementation of an

Table 7.4 Self-assessment audit checklist

AUDIT OF SAFETY POLICY

	Fully met (Score 2)	Partially met (Score 1)	Not met at all (Score 0)
1 The company has a clear, written policy for health and safety at work, signed, dated and communicated to all employees?			
2 The Directors regard health and safety of employees as an important business objective?			
3 The Directors are committed to continuous improvement in health and safety (reducing the number of injuries, cases of work-related ill health, absences from work and accidental loss)?			
4 A named director or senior manager has been given overall responsibility for implementing our health and safety policy?			
5 Our policy commits the directors to preparing regular health and safety improvement plans and regularly reviewing the operation of our health and safety policy?			
6 Our policy encourages the involvement of employees and safety representatives in the health and safety effort?			
7 Our policy includes a commitment to ensuring that all employees are competent to do their jobs safely and without risks to health?			

AUDIT OF ORGANISATION FOR SAFETY CONTROL

	Fully met (Score 2)	Partially met (Score 1)	Not met at all (Score 0)
1 In our company responsibilities for all aspects of health and safety have been defined and allocated to our managers, supervisors and team leaders?			
2 Our managers, supervisors and team leaders accept their responsibilities for health and safety and have the time and resources to fulfil them?			
3 Our managers, supervisors and team leaders know what they have to do to fulfil their responsibilities and how they will be held accountable?			
4 We have identified the people responsible for particular health and safety jobs including those requiring special expertise (e.g. our health and safety advisor)?			

AUDIT OF ORGNISATION OF SAFETY COMPETENCE

	Fully met (Score 2)	Partially met (Score 1)	Not met at all (Score 0)
1 We have assessed the experience, knowledge and skills needed to carry out all tasks safely?			
2 We have a system for ensuring that all our employees, including managers, supervisors and temporary staff, are adequately instructed and trained?			
3 We have a system for ensuring that people doing particularly hazardous work have the necessary training, experience and other qualities to carry out the work safely?			
4 We have arrangements for gaining access to specialist advice and help when we need it?			

Table 7.4 (continued)

5 We have systems for ensuring that competence needs are identified and met whenever we take on new employees, promote or transfer people or when people take on new health and safety responsibilities e.g. when we restructure or reorganise?

AUDIT OF ORGANISATION OF WORKFORCE SAFETY INVOLVEMENT

	Fully met (Score 2)	Partially met (Score 1)	Not met at all (Score 0)

1 We consult our employees and employee safety representative on all issues that affect health and safety at work?
2 We have an active health and safety committee that is chaired by the appropriate director or senior manager and on which employees from all departments are represented?
3 We involve the workforce in preparing health and safety improvement plans, reviewing our health and safety performance, undertaking risk assessments, preparing safety-related rules and procedures, investigating incidents and problem solving?
4 We have arrangements for co-operating and co-ordinating with contractors and employment agencies whose employees work on our site on health and safety matters?

AUDIT OF ORGANISATION OF SAFETY COMMUNICATION

	Fully met (Score 2)	Partially met (Score 1)	Not met at all (Score 0)

1 We discuss health and safety regularly and health and safety is on the agenda of management meetings and briefings?
2 We provide clear information about the hazards and risks and about the risk-control measures and safe systems of work to people working on our site (which is easily accessible in the relevant work area)?
3 Our directors, managers and supervisors are open and approachable on health and safety issues and encourage their staff to discuss health and safety matters?
4 Our directors, managers and team leaders communicate their commitment to health and safety through their behaviour and by always setting a good example?

AUDIT OF SAFETY PLANNING AND IMPLEMENTATION

	Fully met (Score 2)	Partially met (Score 1)	Not met at all (Score 0)

1 We have a system for identifying hazards, assessing risks and deciding how they can be eliminated or controlled?
2 We have a system for planning and scheduling health and safety improvement measures and for prioritising their implementation depending on the nature and level of risk?
3 We have arrangements for agreeing measurable health and safety improvement targets with our managers and supervisors?

4 Our arrangements for purchasing premises, plant, equipment and raw materials and for supplying our products take health and safety into account at the appropriate stage, before implementation of the plan or activity?

5 We take proper account of health and safety issues when we design processes, equipment, procedures, systems of work and tasks?

6 We have health and safety rules and procedures covering the significant risks that arise in our day-to-day work activities including normal production, foreseeable abnormal situations and maintenance work?

7 We have procedures for dealing with serious and imminent dangers and emergencies?

8 We set standards against which we can measure our health and safety performance?

AUDIT OF SAFETY PERFORMANCE MEASUREMENT

	Fully met (Score 2)	Partially met (Score 1)	Not met at all (Score 0)
1 We have arrangements for monitoring progress with the implementation of our health and safety improvement plans and for measuring the extent to which the targets and objectives set under those plans have been achieved?			
2 We have arrangements for active monitoring that involve checking to ensure that our control measures are working properly, our health and safety rules and procedures are being followed and the health and safety standards we have set for ourselves are being met?			
3 We have arrangements for reporting and investigating accidents, incidents, near misses and hazardous situations?			
4 Where the arrangements in 2 and 3 above show that controls have not worked properly, our health and safety rules or procedures have not been followed correctly or our safety standards have not been met we have systems for identifying why performance was substandard?			
5 We have arrangements for dealing effectively with situations that have created risk with priority being given where the risks are greatest?			
6 We have arrangements for analysing the causes of any potentially serious events so as to identify the underlying root causes including causes arising from shortcomings in our safety management system and safety culture			

AUDITING & REVIEWING SAFETY

	Fully met (Score 2)	Partially met (Score 1)	Not met at all (Score 0)
1 We have regular audits of our safety management system carried out by competent external auditors or competent auditors employed by our company who are independent of the department they are auditing?			
2 We use the information from performance monitoring and audits to review the operation of our safety management system and our safety performance?			

Table 7.4 (continued)

---

3 We regularly review how well we have met the objectives in
our health and safety improvement plans and whether we have
met them in the agreed timescales?
4 We analyse the information from performance measurement
and use it to identify future improvement targets and to identify
particular causes of accident, ill health or poor control of risk to
target for future risk-reduction effort?
5 We benchmark the performance of our safety management
system against that of other businesses in the same industrial
sector and/or to monitor our own overall improvement over
time?

---

organisation-wide occupational health and safety management system: a
'Plan, Do, Check, Act' approach is required. Indeed the development and
implementation of a project-based approach to health and safety management is a constituent part of a construction organisation's health and safety
management system.

Within the UK, the development and implementation of construction
project safety management systems is greatly informed by the Construction
(Design and Management) Regulations 2007.

Project Safety Management and the Construction (Design and Management) Regulations 2007

The Construction Design and Management Regulations 2007 (CDM Regulations 2007) are UK regulations that seek to integrate health and safety
into the management of the project. They serve to encourage everyone
involved in a project to work together to:

- Improve the planning and management of projects from the very start;
- Identify risks early on;
- Target effort where it can do the most good in terms of health and safety; and
- Discourage unnecessary bureaucracy (HSE 2007)

The regulations define construction projects as being either 'notifiable' projects or 'non-notifiable' projects. Notifiable projects are those delivered for a
non-domestic client with a construction phase duration that is expected to
be greater than 30 days or 500 person days of construction work. The HSE must
be informed of a notifiable construction project prior to the commencement of
construction phase activity on site.

The CDM Regulations 2007 place duties upon parties involved in both
notifiable and non-notifiable construction projects. These duties are outlined
in *Table 7.3*.

Notifiable projects require a CDM co-ordinator and a principal contractor to be appointed. The production of management information is also required in the form of 'pre-construction information', produced by the client, and a 'construction phase plan' developed and implemented by the principal contractor.

Pre-construction information is project-specific information provided by the client and designers to constructors. This is done in order to assist with both identifying and planning for significant health and safety hazards and risks associated with project design and construction. In undertaking to provide pre-construction information a number of issues must be addressed and documented; these include:

- A description of the project and specifically any information concerning the proposed use of the structure as a workplace;
- The client's considerations and management requirements;
- Any environmental restrictions and existing on-site risks;
- Any significant design and construction hazards;
- Information appertaining to the format of the health and safety file.

(Howarth and Watson 2008)

The approved code of practice that accompanies the CDM Regulations 2007 provides detail regarding the requirements of pre-construction information. These requirements are outlined in *Box 7.1*.

The pre-construction information serves to inform the development of a principal contractor's construction phase health and safety plan. The construction phase plan is required to be developed by the principal contractor before construction work commences. The plan details how health and safety is to be managed throughout the construction phase of a project and is updated and reviewed periodically as required. The approved code of practice that accompanies the CDM Regulations 2007 provides guidance regarding the plan and its contents. This is summarised in *Box 7.2*.

The construction phase plan is a key component of the construction project safety management system. It demands safety management planning to be undertaken and documented prior to construction activity. This documented plan facilitates the co-ordinated implementation and ongoing review of a project-specific safety management system for the construction phase of a notifiable project. Key components of a principal contractor's construction project safety management system are indicated in *Figure 7.9*.

Examples of construction project safety management documentation

A useful tool for reviewing the compliance of a construction project with the CDM Regulations 2007 is provided by the Specialist Engineering Contractors (SEC) Group. The SEC Group have developed a 'Safe

Box 7.1 Pre-construction information requirements

1. Description of project:

    (a) project description and programme details including:

        i. key dates (including planned start and finish of the construction phase); and
        ii. the minimum time to be allowed between appointment of the Principal Contractor and instruction to commence work on site;

    (b) details of client, designers, co-ordinator and other consultants;
    (c) extent and location of existing records and plans.

2. Client's considerations and management requirements:

    (a) arrangements for:

        i.   planning for and managing the construction work, including any health and safety goals for the project;
        ii. communication and liaison between client and others;
        iii. security of the site;
        iv. welfare provision;

    (b) requirements relating to the health and safety of the client's employees or customers or those involved in the project such as:

        i.   Site hoarding requirements,
        ii.   Site transport arrangements or vehicle movement restrictions,
        iii. Client permit-to-work systems,
        iv.   Fire precautions,
        v.   Emergency procedures and means of escape,
        vi.   'no-go' areas or other authorisation requirements for those involved in the project,
        vii. Any areas the client has designated as confined spaces,
        viii. Smoking and parking restrictions.

3. Environmental restrictions and existing on-site risks:

    a) Safety hazards, including:

        i.   boundaries and access, including temporary access – for example narrow streets, lack of parking, turning or storage space;
        ii. any restrictions on deliveries or waste collection or storage,
        iii. adjacent land uses – for example schools, railway lines or busy roads;
        iv. existing storage of hazardous materials;

v. location of existing services particularly those that are concealed – water, electricity, gas, etc.;

vi. ground conditions, underground structures or water courses where this might affect the safe use of plant, for example cranes, or the safety of groundworks;

vii. information about existing structures – stability, structural form, fragile or hazardous materials, anchorage points for fall arrest systems (particularly where demolition is involved);

viii. previous structural modifications, including weakening or strengthening of the structure (particularly where demolition is involved);

ix. fire damage, ground shrinkage, movement or poor maintenance which may have adversely affected the structure;

x. any difficulties relating to plant and equipment in the premises, such as overhead gantries whose height restricts access;

xi. health and safety information contained in earlier design, construction or 'as-built' drawings, such as details of prestressed or post-tensioned structures.

b) Health hazards, including:

i. asbestos, including results of surveys (particularly where demolition is involved),

ii. existing storage of hazardous materials,

iii. contaminated land, including results of surveys,

iv. existing structures containing hazardous materials,

v. health risks arising from client's activities.

4. Significant design and construction hazards:

a) significant design assumptions and suggested work methods, sequences or other control measures,

b) arrangements for co-ordination of ongoing design work and handling design changes,

c) information on significant risks identified during design,

d) materials requiring particular precautions.

5. The health and safety file:
Description of its format and any conditions relating to its content.

(Extracted from Appendix 2 of *Managing Health and Safety in Construction. Construction (Design and Management) Regulations 2007. Approved Code of Practice.* L144 HSE Books)

Box 7.2 The construction phase plan

1. Description of project:

    (a) project description and programme details including any key dates
    (b) details of client, CDM co-ordinator, designers, principal contractor and other consultants
    (c) extent and location of existing records and plans which are relevant to health and safety on site, including information about existing structures when appropriate

2. Management of the work:

    (a) management structure and responsibilities
    (b) health and safety goals for the project and arrangements for monitoring and review of health and safety performance
    (c) arrangements for:

        i. regular liaison between parties on site
        ii. consultation with the workforce
        iii. the exchange of design information between the client, designers, CDM co-ordinator and contractors on site
        iv. handling design changes during the project
        v. the selection and control of contractors
        vi. the exchange of health and safety information between contractors
        vii. site security
        viii. site induction
        ix. on-site training
        x. welfare facilities and first aid
        xi. the reporting and investigation of accidents and incidents including near misses
        xii. the production and approval of risk assessments and written systems of work

    (d) site rules
    (e) fire and emergency procedures

3. Arrangements for controlling significant site risks:

    (a) Safety risks, including:

        i. delivery and removal of materials (including waste) and work equipment, taking account of any risks to the public, for example during access to or egress from the site

    ii. dealing with services – water, electricity and gas, including overhead power lines and temporary electrical installations

    iii. accommodating adjacent land use

    iv. stability of structures whilst carrying out construction work, including temporary structures and existing unstable structures

    v. preventing falls

    vi. work with or near fragile materials

    vii. control of lifting operations

    viii. the maintenance of plant and equipment

    ix. Work on excavations and work where there are poor ground conditions

    x. Work on wells, underground earthworks and tunnels

    xi. Work on or near water where there is a risk of drowning

    xii. Work involving diving

    xiii. Work in a caisson or compressed air working

    xiv. Work involving explosives

    xv. traffic routes and segregation of vehicles and pedestrians

    xvi. storage of materials (particularly hazardous materials) and work equipment

    xvii. any other significant safety risks

(b) health risks, including:

    i. the removal of asbestos

    ii. dealing with contaminated land

    iii. manual handling

    iv. use of hazardous substances, particularly where there is a need for health monitoring

    v. reducing noise and vibration

    vi. work with ionising radiation

    vii. exposure to UV radiation (from the sun)

    viii. any other significant health risks

4. The health and safety file:

(a) layout and format

(b) arrangements for the collection and gathering of information

(c) storage of information

(Extracted from Appendix 3 of Managing health and safety in construction. Construction (Design and Management) Regulations 2007. Approved Code of Practice. P. 106. HSE Books.

Site Access Certificate' which is aimed at helping principal contractors and specialist engineering contractors to comply with the CDM Regulations 2007. The SEC Group state that use of the Certificate facilitates compliance by:

- establishing a clear line of communication and mutually agreed criteria for site safety before the work starts;
- helping to make the work safer by reducing, or removing altogether, the risks arising from poor conditions on site;
- providing a consistent approach to site safety through helping all parties to meet their health and safety responsibilities.

The SEC Group recommend that the checklist certificate is completed jointly by contractors and the principal contractor. A copy of the SEC Group's Safe Site Access Certificate is provided in *Table 7.5*.

Example of a construction project inspection report form

The implementation of the 'Do', 'Check' and 'Act' elements of a construction project's safety management system requires the principal contractor to continuously review and audit health and safety related documentation, communications and the physical site environment. It is good practice to document these audit inspections. *Table 7.6* presents a 'project health and safety inspection report form' for use when documenting a review of a construction project's safety documentation, communication and the site environment.

In undertaking a review of the site environment, regular safety walk inspections enable a review of workplace activities and procedures and any hazards presented in and around the site.

Summary

This chapter has introduced occupational health and safety management systems and has outlined the essential components of such systems. Advocated benefits and problems associated with occupational health and safety management systems have been highlighted and differing standards and guidance documents have been introduced.

The Construction Design and Management Regulations 2007 have been outlined as a means to facilitate a robust systematic health and safety management approach to construction projects. Finally, some examples of useful documentation for supporting the systematic management of health and safety on construction projects have been presented.

Table 7.5 Safe site access certificate

Safe Site Access Certificate
Contract information

The SEC Group (Revision November 2008)

Contract Name: **Principal Contractor:** **Contractor:**	**Brief details of the contractor's work:**
Site Address:	**Areas of the site the contractor will work in:**
Site confirmed as safe and suitable for work by the principal contractor Position: Date: Time: Name: Signature:	Site accepted as safe and suitable for work by the contractor *This must be re-checked at the time of starting work on the site – see page 4.* Name: Position: Signature: Date: Time:

Provision of Information

	No. Weeks	No. Days	
1. Minimum amount of time before start of construction / installation for planning and preparation [REG.22 (1) (f)]	☐	☐ ▲	

	NO	YES	
2. Has the Principal Contractor issued the part(s) of the *construction phase plan* relevant to the work to be carried out? [REG.22 (1) (g)]	☐	☐ ▲	**Details / Comments (in general terms)**

General hazards [REG.22 (1) (i)]

	N/A	NO	YES	
3. Has the Principal Contractor reported the known significant hazards to the Contractor? (e.g. presence of asbestos containing materials, etc)	☐	☐	☐ ▲	**Details / Comments (in general terms)**

Table 7.5 (continued)

	N/A	NO	YES	Details / Comments (in general terms)
4. Has the Principal Contractor given details and locations of all fixed site hazards to Contractor? (e.g. deep water, microwave dishes, contaminated ground, etc)	☐	☐	☐ ▲	
5. Have other suspected or possible significant hazards been advised to the Contractor? (e.g. work by other contractors, such as lifting operations)	☐	☐	☐ ▲	
6. Are there any other site-specific hazards?	☐	☐	☐ ▲	
Site access & storage				
7. Is there clear, adequate and safe access to areas where the Contractor has to work? [REGS 26 & 27] (i.e. free from slipping, tripping and falling hazards, etc)	☐	☐	☐ ▲	
8. Has the Principal Contractor supplied suitable and sufficient site access lighting and power supplies? [REG 44]	☐	☐	☐ ▲	

	N/A	NO	YES	Details / Comments (in general terms)
9. Are the emergency escape routes clear, suitably marked and provided with emergency lighting where necessary? (i.e. a minimum of 5 lux of lighting from battery operated units) [REG 40 (3)]	☐	☐	☐ ▲	
10. Have overhead and underground services and/or obstructions on the site been identified and marked? (e.g. cables, manholes, voids, etc affecting access routes, etc) [REG 34]	☐	☐	☐ ▲	
11. Is hard standing and space available for the delivery and off-loading of huttage, materials, etc, together with easily reached secure storage for materials and/or equipment? [REG.22 (1)]	☐	☐	☐ ▲	
Contractor's personnel				
12. Has a site induction, on site-specific health and safety matters, been given / arranged, before work starts? (This must include particular risks associated with the site) [REG.24]	☐	☐	☐ ▲	
13. Does the Contractor have details of, and understand, the emergency alarms, evacuation procedures and the use of the emergency equipment and services? [REGS.13 (7) & 22 (1)]	☐	☐	☐ ▲	

Table 7.5 (continued)

14. Has the Principal Contractor provided work and rest rooms that are suitable, clean and properly maintained? (i.e. with good room heating, ventilation and facilities) [REGS.13(7) & 22(1)]

☐   ☐   ▲

15. Are the welfare facilities clean, hygienic and properly maintained? (i.e. meeting the minimum regulatory requirements) [REGS.13 (7) & 22 (1)]

☐   ☐   ▲

General protection [REGS. 22 (1) & 26]

	N/A	NO	YES	

16. Are there adequate and effective means of keeping the area/s where the contractor will be working free from:

- other tradesmen and any hazards arising from their work   ☐   ☐   ☐   ▲
- moving plant and vehicles   ☐   ☐   ☐   ▲
- persons using nearby site access routes   ☐   ☐   ☐   ▲
- members of the public and/or visitors   ☐   ☐   ☐   ▲

Note: This can be achieved by physical distance, protective measures to ensure separation (such as a screen), or programming (to separate an adjacent activity in 'time's)

Details / Comments (in general terms)

Other site safety issues [REG.22 (1) (i) (ii)]
17. Are there any other site safety issues that may affect the work?
(These should be listed here or on an attached sheet)

Principal contractor's directions?
18. Are there any specific directions from the principal contractor?
[REG.22(1)] (**These should be listed here or on an attached sheet and cross-referenced to the relevant regulation.**)

Any other comments

Site re-checked at the time of starting work on the site and acceptable
Name: Position:
Signature: Date: Time:

Table 7.6 Construction project inspection report form

Name of Project:	Distribution:
	- Site Manager
	- H&S Manager
	- Commercial Manager
Visit Date:	- Project H&S Committee

Part A: SAFETY DOCUMENTATION & COMMUNICATION

		Y/N	Score 1 – 3*	Comments / Action
1. Safety notices	• A noticeboard clearly displayed for all to see? • Company safety policy? • Site rules displayed? • Appointed safety staff displayed? • Emergency procedures? • Employers' liability cover insurance certificate? • Notification of project to HSE (form F10)?			
2. Construction phase plan	• Available on site? • Up to date? • Management structure and responsibilities? • Emergency procedures? • Fire safety plan?			
3. Method statements & risk Assessments	• Are safety risk assessments completed? • Are risk assessments completed for project scope? • Are method statements available & approved? • Evidence that method statements have been communicated to personnel?			
4. Site manager's weekly inspection reports	• Are these available? • Have actions been addressed (closed out)?			
5. Current report of H&S manager	• Is this available? • Have actions been addressed (closed out)?			
6. Minutes of the site safety committee	• Are these available? • Have actions been addressed?			
7. Records of inductions, plant and permits to work	• Are induction records available and up to date? • Are copies of plant maintenance records available and up to date? • Is a permit to work log / record available and up to date?			
8. Record of accidents & incidents	• Have accidents and incidents been recorded appropriately?			

Part B: THE SITE ENVIRONMENT

		Y/N	Score 1 – 3*	Comments / Action
9. Site security	• Is there a sufficiency of suitable fencing and security on the site to prevent unauthorised access?			
10. Welfare facilities	• Are suitable and sufficient welfare facilities provided in accordance with CDM 2007? • Are the facilities clean? • Are the facilities regularly inspected & maintained?			
11. Materials storage & housekeeping	• Are materials stored in an organised and safe manner? • Is the site tidy of rubbish? • Are recycling bins provided for waste materials? • Is suitable signage placed appropriately around the site? • Are there fire escape routes, assembly points and fire alarms?			
12. PPE	• Is correct PPE worn for all site activities?			
13. Protecting the public	• Are the public protected from: • Site traffic • Falling material • Noise / dust / mud			
14. Pedestrian & traffic routes	• Are there separate vehicle & pedestrian access & egress routes? • Are vehicles and pedestrian routes suitably segregated on site?			
15. Site hazards	• Are the risks caused by specific site hazards and work activities being managed with the application of suitable control measures?			
16. Interviews with site personnel	• Are they CSCS card holders? • Have they received a site induction? • Are they aware of and working to an approved method statement? • Do they receive regular tool box talks? • Do they have any health and safety concerns?			

Score: 3 – Exceeds requirement (excellent); **2** – Meets requirement (good)
1 – Below requirement (poor)

Howarth and Watson (2008)

Questions for the reader

Here follows a number of questions related specifically to the information presented within this chapter. Try to attempt each question without reference to the chapter in order to assess how much you have learned. The answers are provided at the end of the book.

Question 1

Identify the benefits associated with the deployment of an OHS management system.

Question 2

The Institution of Occupational Safety and Health (IOSH) outline a process for the development within an organisation of an occupational health and safety management system. Six typical inputs are identified within this process. Identify these six typical inputs.

Question 3

The Construction (Design and Management) Regulations 2007 place a number of health and safety management duties upon principal contractors when engaging in notifiable construction projects. Identify these duties.

Further reading

European Agency for Safety and Health at Work (2002).*The use of occupational safety and health management systems in the member states of the European Union: experiences at company level.* Available at http://osha.europa.eu/en/publications/reports/307. Accessed 8 May 2010.

European Agency for Safety and Health at Work (2004) *Systems and Programmes Achieving Better Safety and Health in Construction: Information Report.* Available at: http://osha.europa.eu/en/publications/reports/314. Accessed 8 May 2010.

HSE (1998) *Managing health and safety – five steps to success, INDG275.* Sudbury: HSE Books, 1998 (reprinted 2008). Available at http://www.hse.gov.uk/pubns/indg275.pdf.

HSE (2007), *Development of working model of how human factors, safety management systems and wider organisational issues fit together.* Research report RR 543:2007 prepared by White Queen Safety Strategies and Environmental Resources Management for HSE London. Available at http://www.hse.gov.uk/research/rrpdf/rr543.pdf.

Laddychuck, Simon (2008) *Paving the Way for World-class Performance,* Journal of Safety Research (39) pp. 143–49. – A case study example of the development and implementation of an integrated health, safety and environmental management system within a large business organisation.

Rowlinson, Steve (ed.) (2004) *Construction Safety Management Systems,* Spon Press, London. – A compendium of research papers concerning construction safety management systems.

References

Baker, P. 2001. *Raise the safety standard.* Horton Kirby: *Works Management.* 54(11):26–28.

BSI, (2007) *Occupational health and safety management systems–requirements,* OHSAS 18001: 2007. London: BSI.

BSI, (2008) *Occupational health and safety management systems – Guidelines for the implementation of* OHSAS 18001: 2007, OHSAS 18002:2008. London: BSI.

BSI, (2008) *Guide to achieving effective occupational health and safety performance*, BS 18004: 2008. London: BSI.

European Agency for Safety and Health at Work (2002). *The Use of Occupational Safety and Health Management Systems in the Member States of the European Union: Experiences at company level.* Luxembourg: Office for Official Publications of the European Communities.

European Agency for Safety and Health at Work (2009) Working Environment Information Paper: *Occupational safety and health and economic performance in small and medium-sized enterprises: a review.* Luxembourg: Office for Official Publications of the European Communities, 2009.

Howarth and Watson (2008) *Construction Safety Management.* Wiley Blackwell. Oxford.

HSE, (1997) *Successful health and safety management*, HSG65 (second edition). Sudbury: HSE Books, 1997.

HSE (2007) *Managing health and safety in construction. Construction (Design and Management) Regulations 2007. Approved Code of Practice.* HSE Books.

Institution of Occupational Safety and Health (2009) *Systems in focus: Guidance on occupational safety and health management systems.* Source available at: http://www.iosh.co.uk/techguide. Accessed 8 May 2010.

International Labour Office (2001) *Guidelines on occupational safety and health management systems*, ILO-OSH 2001 Geneva.

Laman, Scott (October 2009) *Building a Consensus*, Quality Progress Magazine. Available online at: http://www.asq.org/quality-progress/2009/10/one-good-idea/building-a-consensus.htm. Accessed 8 May 2010.

SEC, (online) *Safe Site Access Certificate.* The Specialist Engineering Contractors Group. Available at http://www.secgroup.org.uk/health.html. Accessed 1 July 2010.

Street (2000:33–35) Getting full value from auditing and metrics. *Occupational Hazards.* 62(8):33–35.

Warren, J., *The role of performance measurement in economic development*, May 2005, AngelouEconomics. Available at: http://www.angeloueconomics.com/measuring_ed.html.

Answers to set questions and case studies

Chapter 1

Question 1

Define the following terms:

1a) Quality policy
1b) Quality objectives
1c) Quality assurance
1d) Quality control
1e) Quality audit
1f) Quality plan

Answers

1a) Quality policy: policy includes the quality objectives, level of quality required by the organisation, and the allocated roles for organisational employees in carrying out policy and ensuring quality. Further it shall be supported and implemented by senior organisational management.

1b) Quality objectives: objectives are a critical component of the quality policy and for example may include establishing the competences required of staff and any associated training, in line with quality policy.

1c) Quality assurance: Kerzner (2001) defined quality assurance as a 'collective term for the formal activities and managerial processes that are planned and undertaken in an attempt to ensure that products and services are delivered at the required quality level'.

1d) Quality control: quality control can be defined as 'a collective term for activities and techniques, within the process, that are intended to create specific quality characteristics'. In other words, it will ensure that the organisation's quality objectives are being met, by using certain techniques such as continually monitoring processes and statistical process control (Kerzner 2001).

1e) Quality audit: this is 'an independent evaluation performed by qualified personnel that ensures that the project is conforming to the project's quality requirements and is following the established quality producers and policies' (Kerzner 2001).

1f) Quality plan: this is a specific quality plan written for a specific project. The plan should contain the key elements/activities of the project and explain in sufficient detail exactly how they are to be delivered and assured.

Question 2

The concept of Total Quality Management has been simplified to four aspects (Haigh and Morris 2001). Identify the four aspects of TQM.

Answer

1. TQM is a total system of quality improvements with decision making based on facts rather than feeling.
2. TQM is not only about the quality of the specific product or service but it is also about everything an organisation does internally to achieve continuous performance improvement.
3. TQM assumes that quality is the outcome of all activities that take place within an organisation, in which all functions and all employees have to participate in the improvement process. In other words an organisation requires both quality systems and a quality culture.
4. TQM is a way of managing an organisation so that every job and every process is carried out right first time, every time. The key to achieving sustainable quality improvement is through the adoption of TQM principles.

Question 3 – case study

You have been asked to act as an external consultant for 'Appleyard Innovative Design Solutions'. Appleyard are considering the implementation of a formal TQM system with a view to obtaining externally verified ISO accreditations. Appleyard consider accreditation to be a necessity in order to be placed on tender lists and continuously improve their operations.

As an external consultant, you are requested to prepare and deliver a presentation to the senior partners of Appleyard. The topic of the presentation is 'the benefits of TQM and the associated implementation process'. Prepare notes to facilitate this presentation.

Answer

TQM can be advocated as a solution for organisations that are under-performing due to their use of traditional organisation structures and management practices whilst operating within a dynamic environment. The implementation of a TQM philosophy can facilitate performance in such organisations.

The advantages of applying a TQM approach are:

- the production of a higher quality product/service through the systematic consideration of a client's requirements;
- a reduction in the overall process/time and costs via the minimisation of potential causes of errors and corrective actions;
- increased efficiency and effectiveness of all personnel with activities focused on customer satisfaction;
- improvement in information flow between all participants through team building and proactive management strategies.

TQM can assist in making effective use of all organisational resources, by developing a culture of continuous improvement. This empowers senior management to maximise their value-added activities and minimise efforts/organisational energy expended on non value-adding activities.

TQM enables companies to fully identify the extent of their operational activities and focus them on customer satisfaction. Part of this service focus is the provision of a significant reduction in costs through the elimination of poor quality in the overall process. This empowers companies to attain a truly sustainable competitive advantage. TQM provides a holistic framework for the operational activities of enterprises. If a firm can overcome the problematic issues of implementation then a sustained competitive advantage is the reward to be gained (Watson and Chileshe 2001).

The TQM implementation process is outlined in *Figure 1.11.*

Chapter 2

Liquidity comments

- The long-term liquidity ratio of 2:1 (the theoretical requirement) is not being achieved, both time periods one approx 1.3:1.
- The acid test shows a worse scenario, the 1:1 is a useful measure. However, the figure for 2010 is only 0.62:1 – i.e. only 62 pence for every £1 of demand.
- This ratio demonstrates that the company had a problem with debt collection in 2009. But the situation has become more critical in 2010. It is taking on average over three months to recover debts.

- The through-put of stocks has improved, thus less capital is tied up in the company.

Profitability

- The profit to capital employed has deteriorated from 2009–10. The company has moved from 10.75 per cent to a loss of 2 pence on every £1 employed.
- The profit generated by every £1 of sales was only 4.76 pence in 2009. However, in 2010 it has reduced to 0.59, thus the company is losing 0.89 on every £1 employed.
- The capital employed is being worked at 2.26 times for both periods.

Table A.1 Case study

Ratio	2009	2010
Liquidity (2:1)		
$\dfrac{\text{Current assets}}{\text{Current liquidity}}$	$\dfrac{1,344,000}{973,000} = 1.38$	$\dfrac{1,464,000}{1,073,000} = 1.36$
$\dfrac{\text{Quick assets (1:1)}}{\text{Current liabilities}}$	$\dfrac{540,000}{973,000} = 0.55$	$\dfrac{663,000}{1,073,000} = 0.62$
$\dfrac{\text{Debtors}}{\text{Sales}} + 365 \text{ (days)}$	$\dfrac{540,000}{2,500,000} \times 365 = \dfrac{78.84}{\text{days}}$	$\dfrac{663,000}{2,230,000} \times 365 = \dfrac{108.51}{\text{days}}$
$\dfrac{\text{Sales}}{\text{Stocks}}$	$\dfrac{2,500,000}{182,000} = 13.74 \text{ times}$	$\dfrac{2,230,000}{90,000} = 24.78 \text{ times}$
Profitability		
$\dfrac{\text{P}}{\text{CE}} \times 100$	$\dfrac{119,000}{107,000} \times 100 = 10.75\%$	$\dfrac{-20,000}{2,230,000} \times 100 = 2.03\%$
$\dfrac{\text{P}}{\text{Sales}} \times 100$	$\dfrac{119,000}{2,500,000} \times 100 = 4.76\%$	$\dfrac{-20,000}{2,230,000} \times 100 = 0.89\%$
$\dfrac{\text{Sales}}{\text{CE}}$	$\dfrac{2,500,000}{1,107,000} = 2.26 \text{ times}$	$\dfrac{2,230,000}{987,000} = 2.26 \text{ times}$

Note: * profit taken as after tax

Chapter 3

Question 1

The construction industry can be divided into five broad sectors where quality assurance is applicable. Identify these sectors.

Answer

- Client in the production of the project brief.
- Designer in the design and specification process.
- Manufacturers in the supply of materials, products and components.
- Contractors (and subcontractors) in construction, supervision and management processes.
- User in the utilisation of the new structure.

Question 2

What are the benefits of ISO 9001:2008 deployment for both the host organisation and its stakeholders?

Answer

- Any initial pressure from customers is inevitably overtaken by the internal energy created by working in a better-managed organisation. Real benefits include:

  - better use of time and resources;
  - improved consistency of service performance and therefore higher levels of customer satisfaction;
  - improved public perception of an organisation's image;
  - enhanced staff morale and job satisfaction: staff understand what is expected of them;
  - transparent and effective communications and processes.

Case study – Question 1

Why may a potential customer demand that a supplier holds certification to ISO 9001:2008?

Answer

Managing an organisation successfully requires a systematic approach. Success can result from implementing and maintaining a management system which is designed to continually improve performance by addressing the needs of all interested parties. Managing an organisation encompasses quality management amongst other management disciplines. Customers are now very sophisticated and they recognise that the deployment of a quality management system is an essential part of a supplier's credentials.

ISO 9001:2008 can demonstrate to a potential customer that the supplier organisation has considered and deployed suitable strategies for addressing the eight key quality management principles of:

- **customer focused organisation** – organisations depend on their customers and therefore should understand current and future customer needs, meet customer requirements and strive to exceed customer expectations. This will provide a valuable assurance to potential customers.
- **leadership** – leaders should establish an organisational unity of purpose, direction and the appropriate internal environment for the organisation, directed at customer satisfaction. They create an environment in which people can become fully involved in achieving the organisation's objectives. One of which has to be satisfying its clients.
- **involvement of people** – people at all levels are the essence of an organisation and their full involvement enables their abilities to be used for the organisation's benefit, and hence for the organisation to meet customer expectations.
- **process approach** – a desired result is achieved more efficiently when related resources and activities are managed as a process matched with customer demands.
- **system approach to management** – identifying, understanding and managing a system of interrelated processes for a given objective contributes to the effectiveness and efficiency of the organisation. Thus a methodological approach is adopted in the delivery of a quality product and or service.
- **continual improvement** – continual improvement is a permanent objective of the organisation.
- **factual approach to decision making** – effective decisions are based on the logical and intuitive analysis of data and information, based upon stakeholder feedback.
- **mutually beneficial supplier relationships** – mutually beneficial relationships between the organisation and its suppliers enhance the ability of both organisations to create value. This value may then be passed on to its customers.

Quality management system requirements as stated in BS EN ISO 9001:2008 not only address the quality assurance of produce and/or service conformity, but also include the need for an organisation to demonstrate its capability to achieve customer satisfaction. This is obviously of vital importance to potential customers.

Case study – Question 2

The Managing Director decided to implement ISO 9001:2008. Why is it essential to have senior management support and what are the likely outcomes if such support is not provided?

Answer

If senior management support is not provided the individual or team charged with the implementation of the quality system is likely to experience problems with regard to:

- inadequate authority to carry the initiative forward and bring it to a successful conclusion;
- insufficient funding for the project, and thus not being able to adequately resource the project;
- insufficient time allocation for the project, thus people do not have the time to contribute;
- resistance to:

  - information and documentation gathering;
  - implementation during the project;
  - maintaining the system.

Successful deployment is dependent upon the strong commitment and involvement of senior management, overtly demonstrated through policies and support.

If companies are to avoid problems relating to resource issues, senior management need to provide the necessary resources. The two most important resource issues are those of adequate funding for the project and the allowance of time for people to participate in the project. Participation is necessary when the quality facilitator/project leader is gathering information for writing of the appropriate documentation. Participation of staff is also vital during the data collection and implementation phase of the project; this fact has to be recognised by and allowed for by senior management.

It should be noted that funding and time allocation are not mutually exclusive. A lack of funds can mean that money is not available to release staff when participation is requested. Also the issues of authority and overcoming resistance to change are not mutually exclusive.

Staff should be delegated sufficient authority to complete their delegated tasks. Senior management should, therefore, make sure that managers are not asked to perform tasks for which they have not been given the necessary authority and/or training.

Chapter 4

Question 1

The European Foundation for Quality Management (EFQM) has stated that the functions of their Excellence Model may be split into four components. Identify these four component parts.

Answer

- as a framework which organisations can use to help them develop their vision and goals for the future, in a tangible and measurable way;
- as a framework which organisations can use to help them identify and understand the systemic nature of their business, the key linkages and cause-and-effect relationships;
- as the basis for the European Quality Award, a process which allows Europe to recognise its most successful organisations and promote them as role models of excellence from which others can learn;
- as a diagnostic tool for assessing the current health of the organisation.

Question 2

The advantages of utilising EFQM EM's self-assessment methodology have been noted by Castka et al. (2003). Identify the advantages of EFQM EM's self-assessment methodology.

Answer

Benefits of using EFQM/self-assessment:

- Providing the opportunity to take a broader view on how the measured activity is impacting on the various business operations.
- Measuring performance of processes, enablers and their relationship with organisational results.
- Self-assessment conducted both internally and externally to the organisation.
- Providing an opportunity to benchmark and compare like for like; or
- Measurement for providing improvement rather than for hard quality control; and
- Self-assessment is also an important communication and planning tool:

  - The results of self-assessment provide a growing common language through which organisations, or parts of organisations, can compare their performances.
  - The outputs of self-assessment are used for strategic management and action planning, or as a basis for an improvement project.
  - New business values: leadership, people, process management, the use of information within the organisation and the way customer relationships are managed.

Question 3

The EFQM EM is based on, and supported by, specific concepts which are referred to as 'The Fundamental Concepts of Excellence'. Identify the Fundamental Concepts of Excellence.

Answer

- Results orientation
- Customer focus
- Leadership and constancy of purpose
- Management by process and facts
- People development and involvement
- Continuous learning, improvement and innovation
- Partnership development
- Corporate social responsibility

Case study

A new managing director has just been appointed to XYZ; the appointment has been made on the understanding that he will oversee the deployment of the EFQM EM within the company. However, the managing director has only a limited knowledge of the model. Yet he has to convince all company personnel of the deployment rationale. Therefore he has decided to engage external consultants to assist him. You have been appointed as external consultant and asked to prepare a presentation for the board of directors (based upon this chapter). Your presentation should consist of key bullet points. The bullet points should relate to the advantages of deploying the EFQM EM; however, you should also note any possible problematic issues of implementation.

Answer

Key benefits of the EFQM model have been recognised:

- It covers all areas of the organisation – offering a holistic approach, which has been absent from many other management approaches that have been used previously.
- It provides for a process of self-assessment against a non-prescriptive but detailed set of criteria, yet is flexible as to when and how this is undertaken. The approach can be adapted to suit the requirements of the user, the size of organisational unit and the extent to which resources can be committed.
- The assessment process is based on factual evidence but the process can be defined at a time and pace to suit the individual organisation. A self-assessment can be completed in as little as a day or with extensive evidence being collected which can take several weeks.
- It offers a means by which other initiatives such as BS EN ISO 9001:2000 can be held and knitted together in an integrated way.
- It offers a way in which a common focus can provide a new way of working that could be embedded into the organisation.

- It provides a balanced set of results indicators, not just financial, that focus on the needs of the customer, the people in the organisation, the local community and other elements of society, the regulatory bodies and the funding providers.
- As the model is used widely across Europe, and has been extensively tested in a range of sectors, private, public and voluntary, it offers benchmarking opportunities with others within and outside the sector, providing a common language to share good practice and develop both individual and organisational learning.
- It provides a framework through which the kernel of the organisation's issues can be exposed, investigated and improved – continually.

The model also engages organisations in an analysis of stakeholders, and particularly supports the recognition of the needs and expectations of customers and customer groups. The EFQM defines customers as the 'final arbiter of the product and service quality, and customer loyalty'. It suggests retention and market-share gain are best optimised through a clear focus on customer needs. In other words it encourages institutions to have a clear focus on the student experience.

The model therefore offers a strong stakeholder-focused approach – which is at the heart of everything. Unless firms are driven by a way of working that looks inside at what is being done and how it is being done for all key stakeholders, then it is unlikely that continual improvement which meets or exceeds stakeholders' expectations could be achieved and sustained.

Chapter 5

Question 1

Illustrate the ISO14001 process in a diagrammatic form.

Answer

Figure 5.1 illustrates the ISO 14001 process in a diagrammatic form.

Question 2

Section 4.2 of ISO14001 requires that an environmental policy must be put into place and be fully supported by top management. What else must top management ensure with regard to the environmental policy?

Answer

An environmental policy must be put into place and this must be fully supported by top management who must ensure the policy:

- is appropriate to the nature, scale and environmental impacts of its activities, products or services;
- includes a commitment to continual improvement and prevention of pollution;
- includes a commitment to comply with relevant environmental legislation and regulations, and with other requirements to which the organisation subscribes;
- provides the framework for setting and reviewing environmental objectives and targets;
- is documented, implemented and maintained and communicated to all employees;
- is available to the public.

(ISO14001 Section 4.2)

Question 3

What are the advocated benefits of ISO 14001?

Answer

A number of benefits can be associated with the implementation of an environmental management system such as ISO 14001.These benefits include:

- assisting an organisation in managing and reducing its impact upon the environment;
- enabling the organisation to meet its corporate social responsibility commitments;
- reducing costs incurred by the organisation due to reductions in, and minimisation of, waste and energy use;
- ensuring that the organisation meets both existing and future environmentally related regulations and requirements;
- increasing the opportunity for inclusion on tender shortlists, as accreditation can be a client prerequisite for this;
- enabling an organisation to obtain a competitive advantage;
- facilitating an increase in confidence in the organisation held by external stakeholders such as financiers and investors;
- enhancing the reputation of the organisation with consumers, clients and potential clients;
- providing a clear basis for the structured training of employees;
- reducing the potential for loss and expense arising from environmental claims due to the considered implementation of control measures and appropriate procedures;
- being a driver for continuous improvement within the organisation.

Question 4

Section 6 of the Site Waste Management Regulations 2008 stipulates the requirements of a site waste management plan. Identify the requirements laid down for a site waste management plan by Section 6 of these regulations.

Answer

Section 6 stipulates the requirements of a site waste management plan:

- A site waste management plan must identify:
  - the client;
  - the principal contractor; and
  - the person who drafted it.
- It must describe the construction work proposed, including:
  - the location of the site; and
  - the estimated cost of the project.
- It must record any decision taken before the site waste management plan was drafted on the nature of the project, its design, construction method or materials employed in order to minimise the quantity of waste produced on site.
- It must:
  - describe each waste type expected to be produced in the course of the project;
  - estimate the quantity of each different waste type expected to be produced; and
  - identify the waste management action proposed for each different waste type, including re-using, recycling, recovery and disposal.
- It must contain a declaration that the client and the principal contractor will take all reasonable steps to ensure that:
  - all waste from the site is dealt with in accordance with the waste duty of care in section 34 of the Environmental Protection Act 1990(3) and the Environmental Protection (Duty of Care) Regulations 1991(4); and
  - materials will be handled efficiently and waste managed appropriately.

Chapter 6

Question 1

Define the terms 'Intrinsic and Extrinsic Motivation'.

Answer

Intrinsic Motivation

This is derived by fulfilling your own needs, and is therefore achieved from work itself. A considerable weight of behavioural scientific research has been devoted to the pursuit of this concept. The importance of providing feedback to employees must be understood and undertaken by managers.

Extrinsic Motivation

This is deriving satisfaction of needs using work as a means to an end. It is sometimes termed the 'instrumental approach'. Work provides us with money and money enables us to 'buy' satisfaction to a certain extent. So pay is the main motivator in this line of thought.

Question 2

Define the advantages of adopting a postmodernist philosophy for a construction company.

Answer

The application of a postmodernist approach to managing companies can provide the following advantages:

- organisations are more flexible and therefore better able to cope with the demands of a changing and challenging work environment;
- the attainment of teamwork and participation at all levels of the company;
- organisational culture is highly motivated and proactive;
- enhanced corporate innovation;
- improved product/service quality;
- a greater market awareness and thus enhanced stakeholder satisfaction.

The identified characteristics of the postmodernist company are essential for an organisation to be able to operate both efficiently and effectively in a dynamic and turbulent operational environment.

Question 3

Define the terms single loop, double loop and triple loop, with regard to organisational learning.

Answer

It can be stated that in 'single-loop learning' people's decisions are based solely upon observations, while in 'double-loop learning' decisions are based on both observation and thinking.

In 'triple-loop learning' a reflection phase is incorporated to support or improve the thinking phase and hence to improve the decision-making process.

'Thus both double and triple loop learning can be considered as generative learning, while single loop learning can be considered an adaptive learning' (Dahlgaard 2004).

Case study

A managing director of a construction company has decided to fully engage with the concept of a learning organisation; however, there exists some resistance to this approach within the host company. The managing director feels that the deployment of the MFAM would be a good starting point for the company. You have been appointed as external consultant and asked to prepare a presentation to convince the rest of the board of directors (based upon this chapter). Your presentation should consist of key bullet points, and these should relate to the advantages of deploying the MFAM.

Answer

Resulting benefits

- A clear understanding of how to deliver value to clients and hence gain a sustainable competitive advantage via operations.
- Enabling the mission and vision statements to be accomplished by building on the strengths of the company.
- Ability to gauge what the organisation is achieving in relation to its planned performance targets.
- Clarity and unity of purpose so that the organisation's personnel can excel and continuously improve.
- Interrelated activities are systematically managed with a holistic approach to decision making.
- People development and involvement. Shared values and a culture of trust, thus encouraging empowerment in line with postmodernist company approaches.

Question 4

The Management Functional Assessment Model (MFAM) provides a focal point for those managers seeking to be proactive in the management of change processes. The MFAM also combines six functions that are usually considered in a disparate fashion, if at all. Identify these six functions.

Answer

- setting and implementing strategic plans;
- setting and implementing operational plans;
- giving due consideration to organisational size, when selecting and engaging in self-assessment linked to an improvement model;
- linking the various functions of management in an effective and efficient way;
- obtaining feedback for stakeholders on organisational performance, with a view to the enhancement of service and product provision;
- building on the concept of triple-loop learning.

Chapter 7

Question 1

Identify the benefits associated with the deployment of an OHS management system.

Answer

Benefits can include:

- Improved prevention of occupational injury and disease – a safer and healthier workplace.
- The provision of a framework for identifying hazards and managing the resultant risks.
- A reduction in the loss of working days due to accidents and injury.
- A reduction in the incidences of employee compensation claims.
- The development of a reviewable approach for meeting legislative requirements, duties of care and due diligence.
- A reduction in insurance premiums.
- Improved morale and productivity brought about by employee inclusivity with developing and running.
- Enhanced working methods that facilitate improvement in production and productivity rates.
- Enhanced reputation of the organisation with a visible and tangible commitment to continuous improvement and inclusive, consultative management mechanism.
- Reduced staff turnover and thereby reduced 'replacement costs'.
- Improved ability to attract skilled personnel.
- Improved commercial potential – inclusion on tender lists is increased as the potential for meeting the pre-qualification requirements of significant clients is enhanced.

Question 2

The Institution of Occupational Safety and Health (IOSH) outline a process for the development within an organisation of an occupational health and safety management system. Six typical inputs are identified within this process. Identify these six typical inputs.

Answer

The six 'typical inputs identified are:

1. Any information relating to hazard identification and risk assessment.
2. Review of OSH performance, including incidents and accidents.
3. Identification and review of existing OSH management arrangements or processes.
4. Competence and training requirements.
5. Workforce involvement.
6. OSH legal and other standards and best practice within sector, e.g. a compliance register

Refer to *Figure 7.8* for further information.

Question 3

The Construction (Design and Management) Regulations 2007 place a number of health and safety management duties upon principal contractors when engaging in notifiable construction projects. Identify these duties.

Answer

The principal contractor's duties are:

- Plan, manage and monitor construction phase in liaison with contractors;
- Prepare, develop and implement a written plan and site rules. (Initial plan completed before the construction phase begins.);
- Give contractors relevant parts of the plan;
- Make sure suitable welfare facilities are provided from the start and maintained throughout the construction phase;
- Check competence of all their appointees;
- Ensure all workers have site inductions and any further information and training needed for the work;
- Consult with the workers;
- Liaise with co-ordinator re ongoing design;
- Secure the site.

Refer to *Table 7.3* for information appertaining to the safety management duties placed upon other parties to a construction project by the Construction (Design and Management) Regulations 2007.

Index